The Pandora Series

XXV

Books in the PANDORA SERIES focus on technology and society – possibilities, risks and uncertainties.

In Greek mythology, Pandora was given a box by the gods but told not to open it. Overcome by curiosity, she opened it anyway. Immediately, all kinds of trials and sufferings flew out over the world. The only thing that Pandora managed to keep in the box was Hope, which is why this has never abandoned humankind.

Editor: Boel Berner

INTERPRETING THE BRAIN IN SOCIETY

Kristofer Hansson
& Markus Idvall (eds.)

Interpreting the brain in society

Cultural reflections on
neuroscientific practices

Arkiv Academic Press

Arkiv Academic Press is an imprint of

Arkiv förlag
Box 1559
SE-221 01 Lund
Sweden

STREET ADDRESS Lilla Gråbrödersgatan 3 c, Lund
PHONE +46 (0) 46 13 39 20

arkiv@arkiv.nu
www.arkiv.nu

A list of Arkiv Academic Press titles can be found in the
last pages of this book. For up-to-date information on
distribution and available titles, please visit:

www.arkivacademicpress.com

Cover design by Lars Jacobsen
Cover photograph by Kristofer Hansson

© The authors/Arkiv förlag
First edition by Arkiv förlag 2017
Arkiv Academic Press international edition 2017
For print information, see the back page of this copy
ISBN: 978 91 980854 9 5
ISSN: 1404-000X

Contents

Preface

Modern neuroscience has a great impact on society, not only on medical treatments but also on existential questions such as how human consciousness can be defined, where feelings arise, when life ends and death occurs. Consequently, current and future neuroscience is a matter of public concern. The argument of this anthology is that such cultural and existential questions should not be left to neuroscientists alone to consider; rather, we should all be engaged in discussing the new knowledge concerning the brain. We suggest in this anthology, perspectives and concepts which may help in creating an understanding of neuroscience and critically scrutinize its various manifestations in society – as a picture of the brain reproduced in a popular science magazine, in how a neurological diagnosis is narrated in a television serial, or as the news of an innovative treatment for those suffering from a neurological disease.

The authors belong to the research group The Cultural Studies Group of Neuroscience, and represent the disciplines of art history, visual studies and ethnology at the Department of Arts and Cultural Sciences, Lund University. The group is in turn part of the multidisciplinary research collaboration BAGADILICO, supported by The Swedish Research Council and which conducts basic research concerning Parkinson's disease and Huntington's disease. BAGADILICO is an acronym constructed from the first two letters in the words Basal Ganglia Disorders Linnaeus Consortium. The basal ganglia is the part of the brain where certain nerve cells die when these two diseases develop in an individual.

We would like to thank all our research colleagues in BAGADILICO, who have made it possible for us to take part in their daily life in the laboratory and study their efforts of involving patients, their families and the public in the conducted research. We have attended a considerable number of their meetings, seminars and gatherings, and made many interviews and observations of their research. Our studies and this anthology would not have been possible without their collaboration. We hope the teamwork will continue! We also wish to express our sincere gratitude to Professor Susanne Lundin, who initiated the cooperation many years ago and has supported us throughout the work.

The result of this anthology would not be the same if not Professor Emerita Boel Berner had given her support in the final process. Her analytical precision has helped us in finalising this book. For this we are very grateful – thank you Boel!

This is not the first book to be published by our group. In 2012, the anthology *The Atomized Body* (edited by Max Liljefors, Susanne Lundin and Andréa Wiszmeg) was published and in 2013, Niclas Hagen defended his doctoral thesis *Modern Genes*.

Kristofer Hansson,
May 2017

Introduction:
Interpreting the brain in society

KRISTOFER HANSSON & MARKUS IDVALL

Society today seems obsessed with the human brain. This is something quite new, since the brain for long was a rather inaccessible and unknown part of our everyday lives. We learned about what the brain looked like and how it worked in biology lessons at school, but apart from that it seemed an irrelevant part of our consciousness and life in general. Recently, however, the brain has come into focus in an entirely new and radical way. It has become a crucial component in our culture, for people's attitudes to themselves and others, and for how they should plan their lives.

A central force behind this cultural change is the new knowledge generated by the expanding neurosciences (Vidal 2009; Pickersgill & Keulen 2011; Rose & Abi-Rached 2013; Schimtz & Höppner 2014). Recent brain research has altered our view of what human beings are and has thus made a large impact on society. This knowledge about the brain has also become a central topic in all kinds of media. It is seen on the internet, referred to in magazines, books, radio and television. Not infrequently, it is the focal point of the plot in television series and films. It seems an inexhaustible source of inspiration for ideas about human abilities, qualities, limits and possibilities. Sometimes, the context is human evolution, and the differences and similarities between the human and the animal brain. In other contexts, the concern is about diseases of the brain and how they contribute to mental and somatic disease, as well as to physical disabilities.

Since the 1960s, the sciences of the brain have been brought together under the discipline of neuroscience. From the start, it was an interdisciplinary science, gathering researchers from a multiplicity of disciplines, such as chemistry, physics, medicine, genetics, and engineering. As one of the world's leading research fields, neuroscience has attracted plentiful resources; it has become a driving force of developments within for example healthcare and medical treatment. Expectations are high of finding a future cure for diseases of the brain that may strike anybody at any time during the life cycle. Older people may suffer from Alzheimer's disease and various forms of dementia. People in mid-life may be afflicted by sudden illness, such as stroke, and among younger people, neuropsychiatric disorders, such as ADHD or Asperger's syndrome, seem to be on an increase.

In this expanding field of knowledge, within neurosciences and in society at large, it has become necessary to take care of the brain as part of our general concern for physical health and wellbeing (Alftberg & Hansson 2012). Physical exercise has become vital, not only to keep the body fit but the brain too, since research results show that the brain is malleable: it responds to what we do and may even be changed through exercise (Doidge 2007). The ageing brain can be vitalized and the growing brain may be stimulated. As if this is not enough, the brain has become the organ of the body that determines if we are considered dead or alive. The concept of brain death has been implemented in many countries, and determines whether a life-saving respirator should be applied in cases of severe brain damage. The brain is also a central reference in our everyday life. Helmets are seen as essential to protect the head (read: brain) against the dangers of modern life: at construction sites, for skiing, when cycling or playing ice-hockey. In sum, the brain is highly present in discourses as well as activities in contemporary society.

The need for critical reflection

In this anthology, we reflect on this emerging knowledge about the brain – how it is understood by scientists and laymen, interpreted in media, and represented in the public domain. We link to previous studies of an expanding neurosciences but add a concern for how participation is expressed, made difficult – or possible – in interactions involving the sciences of the brain. In common is an ambition to *interpret, understand* and *problematise* neuroscientific practices and their impact on human interaction and culture.

We will analyse the negotiations that occur when 'spheres' that are apparently different – such as the scientific community and the public –

interact with each other. What happens when society becomes oriented towards cerebral issues, with a culture that has an increasing element of 'neuro' in its expression and contents? In which ways are groups outside the discipline of neuroscience involved in scientific practices? What existential questions are raised by them, and by scientists, concerning the new knowledge of the brain?

The chapters are written by scholars from art history, visual studies, and ethnology, all involved in a research collaboration with medical and natural scientists who do basic research on Parkinson's disease and Huntington's disease. There is also an afterword by one of these neuroscientists.

The anthology starts with Kristofer Hansson's chapter 'A different kind of engagement: P.C. Jersild's novel *A Living Soul*'. Starting from an analysis of the Swedish novelist Per Christian Jersild's novel *A Living Soul* about the thoughts and experiences of a disembodied brain in a laboratory, Hansson introduces some of the central concepts and perspectives used in the anthology. Issues of participation, ethics and involvement are in focus, as are the existential questions that neuroscientific research often entails.

Many such issues find their way into popular culture. Peter Bengtsen and Ellen Suneson, in their chapter 'Pathological creativity: How popular media connect neurological disease and creative practices', discuss how a dated notion of creativity as linked to neurological disease permeates popular media representation. They give an insightful analysis of two television series – one fictional, one documentary. The role that these programmes play in informing the general public is discussed, and whether this will enable laymen to critically participate in discussions about neuroscientific practices and results.

The next chapter, '"Biospace": Metaphors of space in microbiological images', analyses visual representations in the public sphere. Max Liljefors juxtaposes the well-known images of the earth taken from outer space with those of brain tissue as seen through the microscope; he shows remarkable aesthetic similarities in how they are composed and presented to the public. He argues that to understand the cultural meanings of micrographs it is not sufficient to examine *how* cells, proteins and other biological elements are made to look, we must also ask *where* the images represent them as existing – they are in, what he sees as, an aesthetically constructed 'biospace'.

Andréa Wiszmeg's chapter 'Diffractions of the foetal cell suspension: Scientific knowledge and value in laboratory work' takes us into the heart of neurological research. It is based on interviews with biomedical scientists involved in research on Parkinson's disease. She experiments with a methodological tool – diffractive analysis – to bring forth their different understandings of the meaning of their work with foetal cell suspension. Her aim is to give a nuanced and complex picture of how scientists value what they do, based on different practices and experiences of creating knowledge. Openness and a humble attitude towards such differences are, she argues, necessary for successful interdisciplinary work.

In the next chapter, we continue into the laboratory. In focus are the feelings of biomedical researchers when confronted with the everyday knowledge of Parkinson patients and their families. Kristofer Hansson's chapter, 'Mixed emotions in the laboratory: When scientific knowledge confronts everyday knowledge', focuses on the complex interactions between scientists and patients – as brain tissue in the test tube, in the form of e-mails and letters from patients and relatives, and as live encounters at conferences and meetings. Hansson follows the scientists as they grapple with the everyday handling of the feeling of hope that their research gives rise to among patients, in media and in society at large.

The neurological patient is even more present in the following chapter, 'Meetings with complexity: Dementia and participation in art educational situations', by Åsa Alftberg and Johanna Rosenqvist. They focus on the creation of meaning within an Art Education project which provides guided tours in Swedish museums for dementia-afflicted audiences. Dementia is regarded as a social death, where the affected person is stripped of his or her former subjectivity. Participation, on the other hand, is grounded in beliefs of social inclusion and meetings between people who are active, fully human, subjects. How can we then understand participation in relation to dementia? Alftberg and Rosenqvist explore the complexities of selfhood and embodiment in relation to dementia, based on an intriguing incident with a visitor at one of these art museum tours.

The final chapter, by Markus Idvall, explores this participation perspective further. In 'Taking part in clinical trials: The therapeutic ethos of patients and public towards experimental cell transplantations', he analyses focus group conversations held with Parkinson patients and the public about participation in potentially controversial clinical trials.

Ethical issues of sham, or placebo, surgery, the enrolment of trial participants and the use of cells from embryos and foetuses were discussed in the focus groups. The chapter details the various different standpoints involved and identifies what Idvall terms a 'therapeutic ethos' among the participants – an important component in how the public understands science and the way patients may become involved in clinical research.

The anthology concludes with two afterwords, one written by Malin Parmar, Professor of cellular neuroscience at Lund University, and one by Aud Sissel Hoel, Professor of media studies and visual culture at Norwegian University of Science and Technology. These two scholars set the various contributions in the anthology within a broader multidisciplinary context. Parmar argues that the way we shape our society with the development of new therapies and regulate their use, practicing new knowledge in moving forward, are questions that belong to us all. Hoel deals with relations between neuroscience and aesthetics, and how this perspective can be used to understand participation of non-experts in the discourses on brain and mind.

References

Alftberg, Åsa & Kristofer Hansson 2012: Self-Care Translated into Practice. *Culture Unbound: Journal of Current Cultural Research*, 4: 415–424.

Doidge, Norman 2007: *The Brain that Changes Itself: Stories of Personal Triumph from the Frontiers of Brain Science*. New York: Viking.

Pickersgill, Martyn & Ira van Keulen 2011: Introduction: Neuroscience, Identity and Society. In: Martyn Pickersgill & Ira van Keulen (eds.) *Sociological Reflections on the Neurosciences* (Advances in Medical Sociology, Volume 13). Bingley: Emerald Group Publishing Limited.

Rose, Nikolas & Joelle M. Abi-Rached 2013: *Neuro: The New Brain Sciences and the Management of the Mind*. Princeton: Princeton University Press.

Schimtz, Sigrid & Grit Höppner 2014: Catching the Brain Today: From Neurofeminism to Gendered Neurocultures. In: Sigrid Schmitz & Gritt Höppner (eds.) *Gendered Neurocultures: Feminist and Queer Perspectives on Current Brain Discourses*. Vienna: Zaglossus.

Vidal, Fernando 2009: Brainhood, Anthropological Figure of Modernity. *History of the Human Sciences*, 22(1): 5–36.

1. A different kind of engagement: P.C. Jersild's novel *A Living Soul*

KRISTOFER HANSSON

Imagine an aquarium in a laboratory with an amputated brain floating in sterilised water. The brain still retains its left eye and both ears; these organs register what is happening outside the aquarium. On the outside, the scientists of the laboratory see the brain as an object on which they carry out experiments. But there are other actors in this environment too, who create a greater sense of belonging for the brain, thus making it into a subject, a participant in the world. Such an actor is a girl for whom the brain develops warm feelings. This situation and the environment around the brain – called Ypsilon – is described in the novel *A Living Soul* from the 1980s by the Swedish author and physician Per Christian Jersild.[1] I will make use of it here as a starting point for a discussion about subjectivity of the human brain and how arguments for subjectivity can be linked to the discussion concerning public and patient engagement in neuroscience today. The story starts in medias res when the reader is acquainted with the laboratory environment as seen by the brain from its position in the aquarium.

Suddenly a girl turns up in the brain's field of vision. She is dressed in laboratory clothes and serves Ypsilon breakfast by sucking up a pink fluid from a bottle into a pipette and then letting it trickle over the brain. The girl stands close to the aquarium looking at the brain as she slowly knocks on the glass of the aquarium – at the same time as the brain feels how '[a] thrill of sensual pleasure runs through the fluid I rest in' (p. 5).[2] Thus, a relationship arises between these two subjects. The

---

1. The book was originally published in Swedish in 1980, with the title *En levande själ*. It was translated into English in 1988. In 2014, a short film based on the book was directed by Henry Moore Selder.
2. All quotes are taken from the English translation (Jersild 1988).

novel then draws the reader into the events of the laboratory in which the aquarium is placed, and into the thoughts of the brain. As the girl develops a relationship with the brain and begins to regard it as a subject, so do we, the readers. We understand this subject through the feelings that awaken in Ypsilon for the girl. We also become involved in the thoughts of the brain about the experiments it is submitted to by the researchers. They come to the laboratory to conduct their experiments or just to examine the brain. To the researchers, the brain is an object, which they own, partly through their actions and partly through the fictive company 'Biochine'. Through the researchers' actions and the thoughts of Ypsilon, the reader realises that the researchers have appropriated the right to submit this brain to ethically doubtful experiments. This is done, even though Ypsilon's intelligence is widely superior to that of the researchers. Thus, the novel forces us to consider the ethical basis of such experiments, but also to question theoretically what a human subject is.

The scientists in the novel do not distinguish between humans and human qualities. They therefore understand the brain as a pure object. The reader, on the other hand, who has developed a relationship with the brain even in the first pages of the novel, will regard the brain as a human subject, and will feel with it. In this way, P.C. Jersild forces us, the readers, to problematise our accustomed understanding of what can be regarded as a human subject and what can be considered a researchable object. We are compelled to reflect upon a number of questions. What are the rights of this brain? Should this brain be listened to as a human being? Or is it just one of many objects to be studied? The novel also raises questions about the actions of the researchers. Why can they not understand the implications of their actions? Why do they not listen to what the brain is trying to communicate, about itself and about the experiments it is subjected to? I will discuss this by focusing on issues of human autonomy, and on how technology has altered our visions of the brain and of what human subjectivity may be like. My purpose is to problematise the prospects offered to human beings concerning how they may be able to be involved in decisions relating to themselves as patients or public. I will also connect this discussion to the more elaborated empirical and theoretical perspectives that are presented later on in the separate chapters of this anthology.

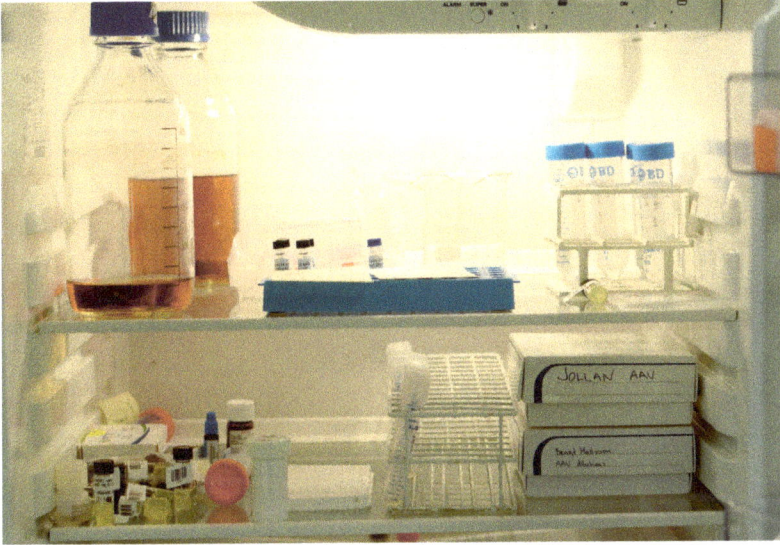

Figure 1.1. Each scientific laboratory has its refrigerator that keeps various fluids of different colours. Photo: Kristofer Hansson.

A free brain?

In our encounter with the experiences of a living brain in an aquarium, as in Jersild's book, human autonomy and subjectivity are called into question. The scientists in the book make matters easy for themselves by defining the brain in the aquarium as an object that can be treated as something non-human. This object, placed in sterile water, does not need to be asked about its views or feelings. It just floats around and can be the object of experiments. But the scientists go further. To be able to use the full capacity of the brain – the purpose is unclear – they try to reset the brain by removing all its memories. One researcher points out: 'In order to give you a fabulous future we were obliged to wash away your past. We had to liberate your cerebral cortex from a mass of old recollections, so that you will be able to assimilate new skills' (p. 77). But what happens with human autonomy and subjectivity when all memories are removed?

In this science fiction narrative, Jersild is saying that this reprogramming and cleansing of old memories of what once was a human being has made Ypsilon into something that is not a subject. His meaning is that Ypsilon can no longer be defined as a full and complete individual, having a right to autonomy and a right to express an opinion about the experiments it is subjected to. From the point of view of the researchers and the company they represent, this appears to be exactly what they intend. Their intention to de-humanize the brain is emphasised in one of the episodes of the book. Here, the researchers do not even call Ypsilon by its name, they just use adjectives and call the brain 'Clever' – an effective way of depriving somebody of their agency and identity. But the scientists go even further when they take an aim at Ypsilon's consciousness as well. 'Don't drive me out [of] my mind' Ypsilon says, terrified of being deprived of the memories it has left, but the researcher says tersely, 'That is exactly what we're trying to do, Clever. It's your mind we need. Not you' (p. 97).

By not giving Ypsilon what we can term full subjectivity, the researchers deprive Ypsilon of the chance of having a say in the matter, of participating in the research. But there are other people in the novel who do develop relationships with the brain, and who regard Ypsilon as a subject. One such relationship is the one between Ypsilon and the girl. Even if there is no body to interact with, the girl tries to understand and respect the brain's thoughts, which it can communicate telepathically due to its immense brain capacity. Ypsilon is here represented as autonomous, since the brain thinks and reflects, no matter what the researchers think and do to it.

Relationships that arise between people – in this case between Ypsilon and the girl, and between Ypsilon and the researchers – are central for how involvement and the chance to express one's opinion are made possible, or become limited. Ypsilon creates a consciousness, an ego through these relationships, even though the researchers simultaneously try to make this ego disappear. From a phenomenological perspective, it could be argued that the researchers cannot get at this autonomy; it exists as an ontological infrastructure, in what the philosopher Jean-Paul Sartre would have termed a dualistic and ahistoric ontology. Ypsilon does not become aware of its self by abstracting itself from the world it is part of; rather, it is this being-in-the-world that perpetuates its consciousness (Sartre 2005 [1943]). Thus, this view of consciousness finds a solution to the body-soul problem by emphasising that humans are aware of themselves through the intertwinement of the body and the ego while being-in-the-world (Merleau-Ponty 2002 [1945]; see also Alftberg &

Rosenqvist's chapter in this anthology). But what happens when there is no body? When the subject is only a brain?

Technology and visuality

In *A Living Soul*, the answer lies in the technology that makes it possible for Ypsilon to be an 'amputated' brain, or bodiless as it were. By depicting this bodiless cyborg, to use Donna Haraway's term (1991), Jersild considers how we might relate to a subject, if the organism of this subject is conjoined to a machine. The brain is both body and soul in Jersild's text. This means that the human subject in the novel is transformed into a kind of 'non-body' where water and glass have replaced flesh and blood. The aquarium essentially becomes a respirator. At the same time, this account raises a question: What is the significance of the conjunction of machine and organism in this encounter?

The novel's description of disconnection from the body and re-connection via technology to the world has been studied from a different perspective in the case of transplantation. When, from the 1960s and onwards, the concept of brain death was introduced within the law in countries with organ transplantation programs, brain function came to define life and death. This replaced respiration and heartbeat and the signs of function of those organs that previously determined whether a person was dead or alive. In this move, the brain became disconnected from the body. Although the body might still be alive and present (the heart might still be beating), death is considered to have occurred since all brain functions have ceased. Then, the respirator is the technology that keeps the body alive even when the person is considered dead. The body exists although the brain is absent, in contrast to the situation in Jersild's novel where the brain is present but the body is absent. The final limit to life – death – is localised to the brain.

One could argue that the concept of brain death produced a new view of what it is to be a human being. Medical anthropologist Margaret Lock posed the following existential and crucial questions in relation to the brain death debate: 'What is a person? What is the relationship of person to body? Does the person cease to exist when the physical body dies? And perhaps the most fundamental, most obdurate question of all: What exactly *is* death – physical, personal and social?' (Lock 2002: 37). The concept of brain death is based in medical expertise and also legitimises this same expertise. Brain death requires surveillance and monitoring of the

Figure 1.2. In the scientific laboratory technology re-connects the biological world and the machinery world. Wires and apparatuses are connected to the organic and vice versa. Photo: Kristofer Hansson.

dying person's body, and availability of technology. Similarly, Ypsilon needs technology, medical expertise, surveillance and monitoring to be kept alive. Technology becomes crucial not only for the relation between life and death, body and soul, and between subject and object; technology also changes our epistemological approach to these dualities. It alters the way we relate to body and brain, to object and subject, and thereby also to how people can be given space for involvement.

More recently, new visual technologies have effectuated altered attitudes to the brain and to what a human being is. Visual representations produced by scientists to depict the brain no longer show only dead brains; illustrative pictures of living brains in activity can also be produced – somewhat like Ypsilon in the aquarium (see Liljefors's chapter in this anthology). Such images are taken with the help of instruments such as Magnetic Resonance Imaging (MRI). MRI and other information technologies are vital for biomedicine's exploration of the brain and its various parts and regions. The technology produces images that must be interpreted by the researchers in order to generate knowledge about the studied phenomenon (Beaulieu 2004). For example, to produce images from a MRI scanner, the machine must for a brief space of time make the hydrogen atoms of the body send out faint radio waves. The waves are detected by the machine and after computer processing, a section image of, for example, the brain is produced. The image is thus not just created in the traditional way by a camera registering objects in front of it; the scanner is itself highly involved in producing the image by means of waves and computer processing. Hence, the knowledge does not only exist in the picture, the image depends on the procedure of the machine processing in the computer and on how its information is interpreted by the observer.

The machine thereby produces new representations of what a brain is (Dumit 2004; Dussauge 2008) and possibly also of the way we might relate to our self. Such visual representations may become a central part of culture and affect how we, as citizens, create space for involvement in this knowledge (Vidal 2009). Consequently, scientists should not be regarded as the sole interpreters of these images; the public, patients and their families may also become interpreters of such pictures, which are published in news or popular media, and are thus available for everyone (see Bengtsen & Suneson's chapter in this anthology).

In a number of texts, media scholar Aud Sissel Hoel has discussed how scientific images change the framing of neuroscientific knowledge (Carusi & Hoel 2014; Hoel & Lindseth 2014). She argues that these new pictures of the brain do not represent facts or visualise a clear bodily phenomenon, rather they are part of an interpretative event, and that interpretation is an important part of our knowledge of the brain. Scientific images produced by the new technology could in such a way be regarded as operative, since they do something to our knowledge. It is no longer only a case of a brain immersed in sterilised water in an aquarium, as in *A Living Soul*. The understanding of the brain also involves visual representations that must be interpreted by experts, both within research and when doctors meet patients, as well as in popular culture. In a chapter, co-authored with philosopher Annamaria Carusi, Hoel argues that when visual representations of the brain are interpreted and understood, a new hybridization occurs between the scientific image and the interpretation made of it. Therefore, the boundary between qualitative medical information and quantitative interpretations becomes indistinct. They write: 'We argue that the blurring of distinctions between qualitative and quantitative methods apparent in this domain in fact challenges our understanding of technologically mediated vision' (Carusi & Hoel 2014: 209). Following Carusi and Hoel, I argue that this is of particular significance when we consider what kind of accountability we as public have in engaging in a rapidly changing scientific landscape.

The complexity of knowledge

The storyline of *A Living Soul* is set within the closed confines of a laboratory; the surrounding world has been shut out. The story unfolds between Ypsilon, the researchers and the girl, together with some other creatures that are also objects of study at the company Biochine. In this

fictional account, Ypsilon's possibility (or lack of possibility) of participating in what is done to it takes the idea to its outer limits. It makes the
reader wonder what kind of place the laboratory is, what day-to-day life
there might be like, and how the scientists may be made accountable for
their actions. The story depicts an all too common notion of research and
of scientists that screen themselves off from the public (Wynne 1996).
Such a fear of the public can be found among biomedical researchers
(Graur 2007) although, in our experience, the case is often the opposite (see my second chapter and Liljefors's chapter in this anthology).
Many biomedical researchers are today highly active in communicating
with patients, their families and the public, inviting them to be part of
a dialogue. Patients are given opportunities to be engaged in the results
of the research and researchers get quite involved in the suffering of the
patients and their families.

P.C. Jersild's novel *A Living Soul* was an early – and fictional – attempt
to raise important issues of scientific knowledge, which thereafter have
become even more urgent and complex. Issues concerning the body, its
diseases and possible cures were by him – and are today – transformed
into existential questions about subjectivity, the individual body and our
selves. Political scientist Herbert Gottweis argues that:

> [a]dvances in human biomedical research (in conjunction with the human
> genome project or new discoveries on the early stages of life) have not only had
> profound and far-reaching practical implications for health care provisioning
> but have also affected our perceptions of what are considered the most basic
> human rights and values (Gottweis 2008: 266).

A similar development can be seen within neuroscience, where the
knowledge generated by research throws existential questions into relief
(Rose & Abi-Rached 2013; Schimtz & Höppner 2014). While this means
that medical science of today can control many bodily processes, it also
entails, according to sociologist Nikolas Rose, that biomedicine and
human life have become politicized; we have here 'politics of life itself'
(Rose 2007). Consequently, these matters cannot simply be left to the
researchers and experts who are a driving force of this development.
Instead, there need to be new ways of involving patients, their families
and the public in questions generated by biomedicine. The main reason
is that there is seldom one answer to these complex questions; rather,
different perspectives provide different kinds of knowledge, to which we
must relate (see Wiszmeg's chapter in this anthology).

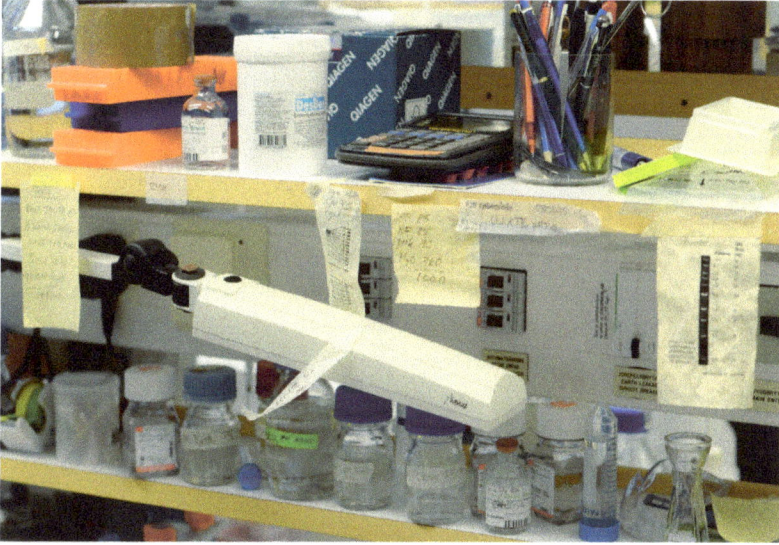

Figure 1.3. Today, knowledge about the brain and its diseases is rapidly evolving. Much of this knowledge originates from the scientific laboratory. Photo: Kristofer Hansson.

Modern biomedicine enables researchers to perform ever more complicated interventions in the brain. These are operations that can be medically risky, but such interventions may also create new ethical dilemmas in regard to limits that should not be transgressed by research. Transplantation of stem cells to the brain is one example. In this case, both the operation itself and the transplantation entails medical risks. But they also involve dilemmas of a more existential nature. Some people are enthusiastic about the prospects opened up by creative biomedicine, while others regard the brain as something humans should not try to manipulate or change (Hyun 2013). Which limits can be stretched by creative research? How can patients, their families and the public be included in discussions about these limits? The matter is complicated by the fact that there is not only a risk involved in innovation, there is also an ethical dilemma involved in not developing medical treatment for severely ill persons (see Idvall's chapter in this anthology).

The involvement of the public is also complicated by what sociologist Luigi Pellizzoni has seen as an imbalance in power relations between science and the public (Pellizzoni 2001). He argues that the scientific community frequently fails in conveying scientific knowledge in a way that is accessible to the public. The knowledge styles used by researchers differ to a great extent from the language and styles of knowledge of the public. Scientists may also use their language and their knowledge style to control who may speak and which arguments are brought up. Thus, communication between researchers and the public is presented by him as being inherently, if not intentionally, restricted by the scientists' use of language and their approach to knowledge. Other researchers maintain that the visualisations showing the results of research are difficult for the public to understand and thereby also the knowledge that is to be conveyed (Pickersgill & Keulen 2011). However, in this anthology we maintain that there are different ways to understand the relation between experts and patients, their families and the public, which hopefully can overcome these dilemmas of different knowledge styles.

Neuroscience thus gives rise to new ethical questions about the risks biomedical research should take or should not take and how they may be understood by the public. There are complex issues in relation to public engagement, to which there is not always a ready answer. There is a necessity to engage patients in the choices that must be made in advance of, for example, clinical tests involving risks. There is also a strong therapeutic promise involved in the developments of science and in what might be possible in the near future, which creates *hope* for those who are afflicted by severe brain diseases (Brown 2003; Rubin 2009). 'You take whatever chances there are', as one person with the neurological Huntington's disease pointed out in an ethnological study (Lundin, Torkelson & Petersen 2016: 72). Issues concerning the involvement of patients and their families in experimental neuroscience have thus become more important, but also more complicated. New knowledge and new questions are generated within brain research but also in regard to how patients, their families and the public view their disease, their brains and their decisions whether to take part in clinical experiments, or not.

Conclusion: interpreting the cultural brain

Within the humanities and social sciences, it has become common to speak about the development of a new neuroscience, which involves other ways of thinking about the brain than the ways we have been used to (Vidal 2009; Williams, Katz & Martin 2011). Therefore, neuroscience and its knowledge is not only something that belongs to scientists; instead it has become an integrated part in our everyday lives. New knowledge and visions about the brain is conveyed to the public through popular media and news media and has become a part in the society. I have in this chapter highlighted some of the questions and theoretical perspectives that may be useful for understanding these neuroscientific practices that are part of our culture today, and in the other chapters in this anthology this is further elaborated.

One central perspective is that new technology produces new representations of what a brain is, and how it can be viewed as an autonomous subject. This knowledge may lead to new possible ways of relating to our self. Not least, the new scientific knowledge about the brain, its diseases and possible cures has in recent decades transformed into existential questions about subjectivity, the individual brain and our self. By problematising our knowledge about the brain and interpreting it in a cultural context, we may also find new public space for engagement in the developments of a variety of neuroscientific practices. Returning to Jersild's novel, it may be that Ypsilon is a lone brain in its aquarium, but it is still a freely thinking subject, part of this world, to which it has a continuous relationship. Or, to put it in Ypsilon's own words:

The entire world is held in the space of a single human brain (p. 9).

References

Beaulieu, Anne 2004: From Brainbank to Database: The Informational Turn in the Study of the Brain. *Studies in History and Philosophy of Science Part C: Studies in History and Philosophy of Biological and Biomedical Sciences*, 35(2): 367–390.

Brown, Nik 2003: Hope Against Hype: Accountability in Biopasts, Presents and Futures. *Science Studies*, 16(2): 3–21.

Carusi, Annamaria & Aud Sissel Hoel 2014: Toward a New Ontology of Scientific Vision. In: Catelijne Coopmans, Janet Vertesi, Michael E. Lynch & Steve Woolgar (eds.) *Representation in Scientific Practice Revisited*. Cambridge: MIT Press.

Dumit, Joseph 2004: *Picturing Personhood: Brain Scans and Diagnostic Identity*. Princeton: Princeton University Press.

Dussauge, Isabelle 2008: *Technomedical Visions: Magnetic Resonance Imaging in 1980s Sweden*. Diss. Stockholm: Kungliga Tekniska högskolan.

Gottweis, Herbert 2008: Participation and the New Governance of Life. *BioSocieties*, 3(3): 265–286.

Graur, Dan 2007: Public Control Could be a Nightmare for Researchers. *Nature*, 450: 1156.

Haraway, Donna J. 1991: *Simians, Cyborgs, and Women: The Reinvention of Nature*. London: Free Association Books.

Hoel, Aud Sissel & Frank Lindseth 2014: Differential Interventions: Images as Operative Tools. In: *The new Everyday: A MediaCommons Project, 'The Operative Image' Cluster*, curated by Ingrid Hoelzl, http://mediacommons.futureofthebook.org/tne/pieces/ differential-interventions-images-operative-tools-2 (accessed 28 February 2017).

Hyun, Insoo 2013: *Bioethics and the Future of Stem Cell Research*. Cambridge: Cambridge University Press.

Jersild, P.C. 1988: *A Living Soul*. Norwich: Norvik Press.

Lock, Margaret 2002: *Twice Dead: Organ Transplants and the Reinvention of Death*. Berkeley: University of California Press.

Lundin, Susanne, Eva Torkelson & Marsanna Petersen 2016: 'With this Disease, you Take Whatever Chances there are' – A Study on Socio-Cultural and Psychological Aspects of Experiments Regarding Huntington's Disease. *Open Journal of Medical Psychology*, 5: 72–87.

Merleau-Ponty, Maurice 2002 [1945]: *Phenomenology of Perception*. London: Routledge.

Pellizzoni, Luigi 2001: The Myth of the Best Argument: Power, Deliberation and Reason. *British Journal of Sociology*, 52(1): 59–86.

Pickersgill, Martyn & Ira van Keulen 2011: Introduction: Neuroscience, Identity and Society. In: Martyn Pickersgill & Ira van Keulen (eds.) *Sociological Reflections on the Neurosciences* (Advances in Medical Sociology, Volume 13). Bingley: Emerald Group Publishing Limited.

Rose, Nikolas 2007: *The Politics of Life Itself: Biomedicine, Power, and Subjectivity in the Twenty-First Century*. Princeton: Princeton University Press.

Rose, Nikolas & Joelle M. Abi-Rached 2013: *Neuro: The New Brain Sciences and the Management of the Mind*. Princeton: Princeton University Press.

Rubin, Beatrix P. 2009: Changing Brains: The Emergence of the Field of Adult Neurogenesis. *BioSocieties*, 4(4): 407–424.

Sartre, Jean-Paul 2005 [1943]: *Being and Nothingness: A Phenomenological Essay on Ontology*. London and New York: Routledge.

Schimtz, Sigrid & Grit Höppner 2014: Catching the Brain Today: From Neurofeminism to Gendered Neurocultures. In: Sigrid Schmitz & Gritt Höppner (eds.) *Gendered Neurocultures: Feminist and Queer Perspectives on Current Brain Discourses*. Vienna: Zaglossus.

Vidal, Fernando 2009: Brainhood, Anthropological Figure of Modernity. *History of the Human Sciences*, 22(1): 5–36.

Williams, Simon J., Stephen Katz & Paul Martin 2011: Neuroscience and Medicalisation: Sociological Reflections on Memory, Medicine and the Brain. In: Martyn Pickersgill & Ira van Keulen (eds.) *Sociological Reflections on the Neurosciences* (Advances in Medical Sociology, Volume 13). Bingley: Emerald Group Publishing Limited.

Wynne, Brian 1996: Misunderstood Misunderstanding: Social Identities and Public Uptake of Science. In: Alan Irwin & Brian Wynne (eds.) *Misunderstanding Science? The Public Reconstruction of Science and Technology*. Cambridge: Cambridge University Press.

2. Pathological creativity: How popular media connect neurological disease and creative practices

PETER BENGTSEN & ELLEN SUNESON

Within neurological research there is an extensive interest in studying how creativity and artistic expression are related to brain activity (e.g. Demarin 2009; Pąchalska et al. 2013; Piechowski-Jozwiak & Bogousslavsky 2013; Zaidel 2014). This interest is also reflected in popular science and fiction narratives related to neuroscience. Popular media play a vital part in the general public's access to, and understanding of, scientific research. Therefore, as communication scholars Matthew Nisbet and Declan Fahy (2013) argue, the narratives media present have a considerable impact on how society takes on board scientific results.

While television viewers have long been presented with portrayals of the world of general medicine in fiction series like *ER* (1994–2009), *House* (2004–2012) and *Grey's Anatomy* (2005–), the developing fields of cognitive neuroscience and neurology have also been represented more specifically in popular media. Taking as our main points of departure the BBC-produced documentary series *Brain Story* (2004) and ABC's one-season fiction television series *Black Box* (2014), we begin this chapter by considering the role these programmes play in informing and enabling the general public to critically participate in discussions about neuroscience.[1] We then go on to discuss the programmes' recurring linking, often through references to art history, of neurological disorder and creativity. We further discuss the implications this linking may have for the dissemination of ideas about personality traits, artistic expression, neurological

---

1. We have chosen these programmes to highlight how the same types of art historical narratives recur in similar ways within both popular science and fictional television programmes.

disease and neuroscience as a discipline. We examine in particular how artistic expressions are used when coming up with diagnoses, and we discuss and problematise the media portrayal of the neurologist as a kind of art critic who – by studying for example the colour, painterly style and motif of artworks – ostensibly is able to gain insight related to the neurological processes in artists' brains, and ultimately categorise them as 'normal' or 'pathological'.

Popular media and potential for participation

As mentioned above, popular media plays an important role in informing and enabling the general public to participate in discussions about neuroscience. The American media scholar Henry Jenkins (2006) argues that when knowledge is presented to the public through for example television programmes, it is mediated – or translated – through genres that move in an intermediate terrain between fact and fiction proper. He also makes the point that presenting research through certain thematic narratives and frames of reference makes complex scientific results appear coherent and accessible. To be relatable and to engage the audience more easily, such narratives are often infused with historical, fictional and cultural tropes that are already well-known to viewers.

While popular media products can help relate complex topics to a broad public in an understandable manner, it is important to note that this is not the same as participation. The latter occurs, rather, when audiences critically engage with the content as part of a process, which Jenkins refers to as *convergence*. Essentially, convergence is a 'word that describes technological, industrial, cultural, and social changes in the ways media circulates within our culture' (Jenkins 2006: 282). Contemporary society is characterised by what Jenkins calls *convergence culture*, which entails people following and to some degree taking control of media content over a number of different platforms and in different contexts, often to the point where the line between producer and consumer becomes blurred.

Jenkins sees user involvement as a key component of convergence. User involvement in convergence culture as described by Jenkins may, for example, take the form of viewers discussing media products with each other on internet forums, researching and fact-checking information presented in the shows and creating new media content like animated GIFs, fan fiction, analyses and reviews based on the original programmes. These secondary, viewer-generated media products may be circulated on a number of platforms, and may in turn be appropriated, reworked, discussed and redistributed. This simultaneous consumption and production of media often takes place within so-called *knowledge communities*. These are social – often online-based – environments that usually 'form around mutual intellectual interests; their members work together to forge new knowledge often in realms where no traditional expertise exists' (p. 20). If picked up by knowledge communities that are able to engage critically with the content, popular media programmes can be efficient vehicles for increasing awareness and fostering public debate about specific themes like neuroscience and neurological disease.

In relation to media products like the popular science programmes that constitute part of the empirical material for this study, however, there seems to be a lacking interest in engaging on this kind of partici-patory level. Convergence seemingly occurs more frequently around programmes that have dramatic narratives and characters people can – and are meant to – become emotionally invested in. This leaves popular science programmes like *Brain Story* at a disadvantage when it comes to viewer engagement and de facto participation. Having said this, it is interesting to note that while *Black Box* is a fiction drama with story arcs and characters that viewers are supposed to become invested in, the convergence and participation connected to this series also seems to be limited. There is little empirical material to suggest the existence of a dedicated fan base that engaged actively with the series or produced sec-ondary media content while, or after, the show was running. Thus, on 28 February 2016, The Cube, a community on TV.com dedicated to *Black Box*, had 503 members. On fanfiction.net, a total of five fan-made stories taking place in the series' universe have been published (all between 3 June 2014 and 22 June 2015). For comparison, the long-running medical drama series *Grey's Anatomy* (2005–) currently has approximately 12 800 fan-made stories uploaded to fanfiction.net. Fifteen of these were posted during the airing of the series' first season. Significantly, each of these fifteen stories has an average of just over 13 reviews with no noteworthy outliers (the lowest number of reviews is 6, the highest 20), indicating an active fan base. The average number of comments is actually higher for the *Black Box* fan fiction (14.4 reviews), but it is worth noting that one of the five stories has received 53 reviews. Without this outlier, the aver-age number of reviews for the rest of the stories is 4.75, which indicates a relatively low level of interaction among fans.

The vast majority of the debate surrounding *Black Box* took place in commentaries related to its cancellation after just one season (the latter also being a sign that viewers were not engaging with the show). A common theme in the responses to the cancellation is that people relate the show's depiction of neurological disease and treatment to their own personal experiences, often to emphasise how realistic *Black Box* was. Several commenters thus describe the show as not just entertaining but also educational, and blame the cancellation on a lack of understanding of the issues the show discussed. One commenter on TV.com wrote:

> They cancelled it because of general audience ignorance. Most seriously intelligent shows don't make it. And this was a very complicated series. I miss it, because of the ground-breaking medical treatments. Incredible show, complicated characters, informative & entertaining.[2]

The mention of 'ground-breaking medical treatments' and the characterisation of the show as 'informative' are interesting, because they seem to be based on the notion that what is presented as facts about the brain, neurological diseases and their treatment in *Black Box* is also valid outside of the universe the show established. At a time when comedy shows like *The Daily Show*, *The Colbert Report* and *Last Week Tonight* are seen as news programmes, this is perhaps not surprising. Previously established genres are converging, and the delimitation between fact and fiction is becoming blurred. However, it is worth noting that in popular media products, nuances tend to disappear and the complexities of neuroscience – for example reading an MRI scan – are glossed over in favour of maintaining a compelling narrative. Without entities like knowledge communities to critically engage with the content, popular media programmes may therefore end up as vehicles for transmitting dramatically effective, but scientifically inaccurate, ideas and stereotypes, such as the existence of a direct link between neurological disease and creativity.

Extraordinary brains and the duality of neurological disease

The ABC-produced fiction television series *Black Box* (2014) has as its main protagonist Dr. Catherine Black, a neurologist who secretly lives with bipolar disorder. The title of the series obviously echoes the pro-

---

2. Comment by 'joycewrandolph' on 26 February 2015. Retrieved from: http://www.tv.com/shows/the-black-box/community/post/black-box-canceled-abc-140745646742/ (accessed 28 February 2016).

tagonist's surname. However, 'black box' is also a term commonly used to describe a system or an object in a situation where one has access to input and the resultant output, but no solid understanding of what goes on inside the object itself. The human brain, in its reception of and response to stimuli, is often said to be precisely such an object.[3]

The series' first episode opens with Black in her psychotherapist's office, recounting her experiences leading up to, during and after giving a keynote speech in San Francisco to the (fictional) Neurological Institute of America. Black explains that she had been preparing for weeks, but felt her manuscript was subpar upon rereading it at her hotel the day before the presentation. She further reveals that she therefore chose to skip her medication in order to enter into a manic state and get what her therapist then refers to as 'a shot of inspiration'. In a flashback, we see the result as Black works through most of the night. She goes on to speak enthusiastically in front of a large audience about her efforts to make sense of the so-called ordinary brain by studying extraordinary ones. As part of her reasoning for this approach to her research, she refers to the Dutch artist Vincent van Gogh (1853–1890), whom she paraphrases as having said that in order to understand the colour blue, it is first necessary to understand its opposites, yellow and orange. Although not made explicit in the television series, this appears to be in reference to a letter from 1888 from van Gogh to the artist Émile Bernard. In it, van Gogh writes:

> What I should like to find out is the effect of an intenser blue in the sky. Fromentin and Gérôme see the soil of the South as colourless, and a lot of people see it like that. My God, yes, if you take some sand in your hand, if you look at it closely, and also water, and also air, they are all colourless, looked at in this way. There is no blue without yellow and without orange, and if you put in blue, then you must put in yellow, and orange too, mustn't you?[4]

Black's description of colours as opposites seems to suggest to the audience by analogy that so-called 'ordinary' and 'extraordinary' brains are also to be seen as opposites rather than variations. As will be demonstrated presently, however, the series soon contradicts this point of view.

---

3. Not all viewers understood the connection between the show's title and the brain as a 'black box'. In the wake of the show's cancellation, one commenter wrote: 'This was an awesome show! I think the title did not do it justice. I had to watch to realize the show was GREAT! It just had a stupid name.' – Comment by 'lindaspinoza9' on 12 March 2015. Retrieved from: http://www.tv.com/shows/the-black-box/community/post/black-box-canceled-abc-140745646742/ (accessed 9 December 2015).
4. Retrieved from: http://www.webexhibits.org/vangogh/letter/18/B06.htm (accessed 26 October 2015).

The conference speech is the first of numerous instances in the series where links are made on different levels between neurology, creativity and art. First, in the scene a parallel is drawn between the methodologies of Black and van Gogh, staging the scientist as creative. This running theme in the series is in part motivated by Black's bipolar disorder, which ostensibly enables her to relate more easily to her patients and gives her an unconventional and creative angle to work from. Second, apart from the reference to understanding things through their opposites, Black also uses van Gogh as her main example of a person with an extraordinary brain. She does so by way of an anecdotal story about the creation of his famous oil painting *The Starry Night* (1889). A projected image of the painting appears in the background during her presentation (see figure 2.1). *The Starry Night* depicts a swirling starry sky sprawling over a small town. It is one of van Gogh's most famous paintings, and was also a key work in what came to be known as Expressionism; a modernist art movement that focused on the use of strong colours and distorted forms in order to express the inner life and feelings of the artist. Later in this text we will elaborate on the frequent connections made in our empirical material between Expressionism and mental illness, which serve to frame neurological research.

Cutting back and forth between the flashback and the conversation with her therapist – scenes that respectively present the protagonist in a manic and medicated state – Black explains that van Gogh was a patient at a mental hospital when he created the painting. She then goes on to list a number of other people she deems to be extraordinary and uses them to problematise the idea and ideal of the normal:

> Temporal lobe epilepsy allowed Saint Paul to hear the voice of God. Hemingway, Sylvia Plath, Billie Holiday, Charles Dickens, Herman Melville. These are just a few of the great minds that suffered from a fine madness. Should they have been medicated into mediocrity? My work is about respecting each and every individual brain, and while I learn from my patients, I make no attempt to distinguish them from any imaginary state of normalcy.[5]

After this speech against the ambition to normalise patients through medication, Black seemingly receives a standing ovation from the audience for her revolutionary take on neurological afflictions and their (lack of a need for) medicinal treatment. In the conversation with her therapist, however, the audience's reception is revealed to be a hallucination. While the series in this sequence on the surface seems to advo-

5. From *Black Box*, season 1, episode 1: 'Kiss the Sky'. First broadcast on 24 April 2014.

Figure 2.1. Dr. Catherine Black in front of a projection of The Starry Night *(1889) by Vincent van Gogh. See* Black Box, *season 1, episode 1: 'Kiss the Sky'. First broadcast on 24 April 2014.*

cate explicitly for the benefits of non-medication, the manic origins and the hallucinated reception of Black's speech purposefully undermines the credibility of such an approach by showing that her non-medicated state is in fact counterproductive.

After delivering her speech, Black returns to the hotel, where her manic state peaks. She runs out onto her room's balcony and does an interpretive dance to a piece of experimental jazz. There is significance to the recurring use of this type of music by the series in scenes where the protagonist is in a manic state, as the improvisational and expressive nature of experimental jazz can be said to underscore, on an aural level, the link between neurological disease and creativity. It is dark on the balcony and to begin with the night sky is lit up only by the moon and by the San Francisco skyline. As Black falls deeper into her manic state, however, the scenery changes; stars begin to swirl in formations across the night sky and the moon radiates light in a way similar to the previously depicted van Gogh painting (see figure 2.2). The scene in this way visualises the connection between Black and van Gogh, which was previously established verbally in the former's speech. Climbing onto the railing, Black leaps off the balcony and proceeds to float among the stars over San Francisco. However, her flight is soon revealed to be another hallucination. Slipping off the railing, Black lands on the balcony floor, narrowly avoiding a fall that could have ended her life.

Figure 2.2. Dr. Catherine Black in front of a San Francisco night scenery that resembles The Starry Night *(1889) by Vincent van Gogh. See* Black Box, *season 1, episode 1: 'Kiss the Sky'. First broadcast on 24 April 2014.*

The balcony scene encapsulates the duality of neurological disease as presented in the series. On the one hand, through the visions, music and dance, the scene visually and audibly communicates the creative potential in experiencing the world differently. On the other hand, Black's hallucinations bring her close to falling off a building, potentially causing serious harm or death to her and possibly to others. This is the edge that we find Black balancing on throughout the series, and it is simultaneously the edge that she repeatedly is tasked with pulling her patients back from.

Framing neuroscience and neurological disease as creative

The way popular science programmes and fiction television series portray neurological disease, cognitive neuroscience and neurology contributes to shaping discourses about the body, including the brain, as well as the scientific disciplines themselves (Dumit 2004; Nisbet & Fahy 2013). Our perception of the human body is not something that is naturally given. Instead, it is the result of an ongoing process where societal, cultural and scientific beliefs interact with, and form, our notions of our bodies and selves. In this continuous reproduction and negotiation of ideas, information coming, at least ostensibly, from science is highly influential. The communication researchers Matthew Nisbet and Ezra M. Markowitz use the concept of *framing* to explain how media coverage and popular discourse often embed scientific results in well-known narratives when presenting them to the general public. They argue that these frames in turn shape public perceptions and opinions about current research and its results, and ultimately affect core beliefs about science and society (2014: 1).

We have found that both fiction and popular science television programmes frequently discuss the relationship between neurological disease, creativity and artistic ability. Interestingly, they do so using a relatively outdated framing of the artist as a sole creative genius. This stands in contrast to more contemporary theories of art production and interpretation, which tend to see the artist as one of a number of agents involved in the production of art. From this point of view, the artist is perceived as an immediate producer, who is part of a larger system of for example suppliers of materials, gallerists, art critics, collectors and other artists. This system, whose members collectively constitute phenomena – material or otherwise – as art, is what the American sociologist Howard S. Becker (1982) calls an *art world*. The French sociologist Pierre Bourdieu (1993) refers to essentially the same social phenomenon as a *field of cultural production*. Since, as mentioned above, the framings of neurological disease as connected to creativity are likely to influence the field of neuroscience and the public's understanding of neurological disease, it is interesting to examine the discourses that are embedded in these specific narratives and explore why, for example, the notion of the artist as a sole genius may be convenient as a part of an explanatory framework in popular neuroscience.

Scholars in fields like anthropology (Dumit 1997 & 2004), sociology (Rose & Abi-Rached 2013) and communication studies (Beaulieu 2000 & 2002) have pointed out how popular media increasingly frame neurological processes as the origin of human behaviour and personality traits. One such trait is creativity. We have found that three predominant narratives are repeated by popular media to frame or discuss neuroscience and neurological disease in relation to this particular attribute. First, as seen in the above description from *Black Box* of van Gogh and his work, neurological disease is commonly used as an explanation for creativity. Second, the framing of a link between neuroscience and creativity is often established by affording the neurologist or neuroscientist the role of a contemporary, science-oriented version of the artistic genius. This is seen in the above empirical example where Black claims to draw inspiration for her own work from the observations of van Gogh, and seemingly comes up with novel perspectives on the nature and purpose of neuroscience as well as on the symptoms of neurological disease. Additionally, Black's own creativity is portrayed as interlinked with her bipolar disorder, thus following a well-known narrative of the artist as suffering, but also benefitting, from mental illness. Third, the neurologist or neuroscientist is presented as a kind of art historian or art critic who is studying artistic expressions in order to determine their neurological origin. This can be seen in several instances in *Black Box*, for example in Black's reading of van Gogh's work. As will be shown in the following section, this framing of the neuroscientist is also found in real-life cases presented in the popular science programme *Brain Story*.

In the above we have briefly presented parts of our empirical material and theoretical perspectives as well as three main narratives related to creativity that are commonly used frames of reference when popular television programmes mediate neuroscientific research to a general audience. In the remainder of the chapter, we will further explore these narratives and connect them to examples from our empirical material.

Neurological disease as an explanation for creativity

In the first episode of the BBC-produced popular science programme *Brain Story* (2004), the host, Professor of pharmacology Susan Greenfield, discusses how Vincent van Gogh's temporal lobe epilepsy may have affected his artistic imagination.[6] In her introduction to the segment on the artist, Greenfield questions whether it could be that van Gogh's 'epilepsy led not only to his crippling mental problems, but at the same time to his awesome creativity'. She does so while sitting in the garden of the mental asylum Saint Remy in southern France, where van Gogh was admitted in 1889. The artist made several paintings of the landscape around the asylum, and Greenfield compares these paintings to the actual surroundings while stating that the comparison poses interesting questions about the way the artist saw the world and how the physical upheaval in his brain somehow transformed his perceptions. This interpretation of works of art, as direct expressions of the artist's self and the artist's particular way of seeing, reflects a view of art derived

6. *Brain Story*, episode 1: 'All in The Mind'. First broadcast on 8 May 2004.

from the Western nineteenth-century concept of Romantic individualism. During the Romantic Age (from around 1790 to 1850) the 'mad' were commonly viewed as superior beings and ascribed an imagination with great transcendental creative powers. In this way, the idea of an affinity between madness and artistic ability was established. Romantic individualism later influenced modernist ideas of subjectivity and art; artists were perceived as individual geniuses, and their artworks were seen as expressions of their particular identity or 'inner world' (Porter 2009 [1999]). As mentioned previously, however, such a perspective is no longer commonplace within the field of contemporary art theory.

The segment about van Gogh in *Brain Story* provides a detailed analysis of his art from the perspective of a neurologist. While the camera zooms out to reveal a self-portrait by van Gogh, and then dwells on a close-up of the portrayed artist's eye, Greenfield's voiceover along with excerpts from an interview with neurologist Shahram Khoshbin account for the way van Gogh's epilepsy may have affected the area of his brain just behind his temples. Furthermore, Khoshbin explains that this part of the brain, the temporal lobe, is where sensory integration between vision and hearing takes place. On this basis, Khoshbin argues that 'it is easy to see how a disturbance in this area could create a different sensory experience'. The programme then cuts to a view of a landscape, presumably the Saint Remy garden, filmed through a window with patterned glass (see figure 2.3). The irregular structure of the glass causes the view of the landscape to become distorted, making it somewhat reminiscent of the paintings by van Gogh. In her voiceover, Greenfield adds to this visualisation the information that 'the epilepsy that van Gogh probably suffered is not uncommon, but in a small number of cases the resulting uncontrolled brain activity can permanently change the way a person perceives the world'. This commentary furthers the idea that van Gogh's perception of the world was heavily influenced by temporal lobe epilepsy, and that this fundamentally affected his work as an artist.[7]

---

7. The idea that van Gogh really did see the world differently is also a prominent plot point in series 5, episode 10 of the BBC-produced science fiction series *Doctor Who*, entitled 'Vincent and the Doctor' (first broadcast on 5 June 2010). In the episode, van Gogh is portrayed as being able to actually see colours, swirls and burst of light in the night sky similar to those depicted in *The Starry Night*. In addition, he can see, and is therefore able to ultimately defeat, a monster that is invisible – but very real – to the other characters. While not directly linked in the show to temporal lobe epilepsy, the *Doctor Who* episode draws on an established narrative of van Gogh's unique perception of the world.

Figure 2.3. A landscape view filmed through a patterned glass window. In Brain Story this is meant to visualise van Gogh's distorted vision. See Brain Story, episode 1: 'All in The Mind'. First broadcast on 8 May 2004.

While it certainly cannot be ruled out that a neurological affliction can influence the way the world is perceived and subsequently depicted, the link between creative output and neurological disease is not as clear-cut as the representation in *Brain Story* might lead us to believe. Other factors should also be considered, such as the prevailing ideas and artistic tendencies of the time. As Nisbet and Fahy (2013) argue, popular media often use tales or metaphors familiar to the general public in order to embed complex scientific results in a more comprehensible framework. While this kind of mediation of results makes otherwise complicated and inaccessible information available to a wider audience, the prominence of one narrative also often means that other possible narratives or perspectives are disregarded. In the case of van Gogh, contemporary art histori-ans have traced the strong influence of his close collaboration with the artists Paul Gauguin (1848–1903) and Émile Bernard (1868–1941). As art historian Anne-Birgitte Fonsmark (2014) has pointed out, the three art-ists' motifs, choice of colours, compositions and other stylistic elements were 'wandering' between them. In other words, van Gogh's artistic out-put cannot be understood simply through the perspective of mental afflic-tion. It was at least in part a result of influences from the social context he was a part of; a time and an environment where styles that later became known as characteristic of art movements such as Post-Impressionism and Expressionism began to emerge. In addition to the social aspects, astronomer Charles A. Whitney (1986) has pointed out that van Gogh would sometimes compose his night skies by combining different vantage points. Art historian Evert van Uitert (2016) further writes that van Gogh's *The Starry Night* was composed 'not directly from nature, but with the help of sketches' and that 'the old, rather brutal woodcuts that illustrate the Household edition of *The Works of Charles Dickens*' served as a source of inspiration for the style of the painting. That the paintings are composites is a further indication that van Gogh did not simply depict the world as he saw it and that the idea of a direct link between neurological pathology and artistic ability is problematic.

As mentioned previously, the narrative of the artist as a specific 'personality type' and of art as expressive of a particular kind of identity has its roots in a historical period within Western thought, which reached its pinnacle in nineteenth and twentieth century European modernism. The highlighting of art as an individual expression was initially connected to the idea of the artist as a sole, divinely inspired genius. This point of view meant that the identity of the artist was collapsed into the artwork. Art thus came to be seen as a reflection of the artist's 'self' (Kester 2011: 156). A search for 'pure' individual expressions unaffected by societal norms meant that some modernist art movements, like Art Brut, took an interest in art made by children and the mentally ill. Other art movements, such as Dadaism or Surrealism, developed automated methods for making art. By doing so, they hoped to access expressions of the subconscious that would otherwise be repressed. As mentioned previously, during the twentieth century, these ideas have become highly contested because they do not take into account the contextual, intersubjective and social aspects that play a central role both in the formation of identity and the development of an artistic style.

Given the rather outdated nature of the modernist narrative of identity, it is interesting that it is recurrently used in popular mediations of neuroscientific research. As a case in point, van Gogh is a widely known art historical character, and as a perceived artistic genius he is commonly used as an example when connections are made between mental illness and creativity. In contrast to contemporary art theory, *Brain Story*'s interpretation of van Gogh's work is similar to the modernist idea of how the identity of the artist is collapsed into the artwork; it differs mainly by a change in focus. In this popular neuroscientific reading, the idea of the artist as being guided by his or her inner life seems to have been supplanted with an idea of the artist as being controlled by his or her brain activity. In the following section, we will further discuss how this relationship between neurological disease and creativity is framed.

Framing artistic ability as a symptom

One of the medical cases handled by Dr. Black in *Black Box* is introduced by showing a young man, named Anthony Guiness, manically creating a painting in strong colours on the wall and doors of a hospital room (see figure 2.4). When neurologist Catherine Black walks in, Guiness' parents, who have been sitting at a table in the room, apologise for their son's behaviour and explain that they are unable to stop him. They further offer to pay to have the room repainted. Black, however, refuses and remarks that the painting is beautiful.

The parents tell Black that their son has always been an exceptional student and that he was headed to MIT to study physics. They further explain that 'he has never been interested in art. He started drawing about three months ago. First he drew all over the walls of his room, and then the school suspended him for defacing the hallways'. When Black points out that the emergency room diagnosed their son with schizophrenia his mother exclaims: 'That is wrong! Our son is a scientist, he is *not* an artist. Something happened to him, something changed in his brain!' This comment simultaneously draws on and feeds into the framing of neurological disease – in this case schizophrenia – as a source of creativity and artistic ability. The mother clearly links being an artist with being mentally ill. This is a similar narrative to that found in the previously discussed segments from both *Black Box* and *Brain Story* on van Gogh's temporal lobe epilepsy.

The case as a whole also allows the series to discuss the viability of using the analysis of creative output to diagnose patients. The problematic nature of this type of diagnosis is demonstrated later in the episode, as it turns out that a brain tumour, not schizophrenia, is causing the patient's change of personality and outbursts of creativity. This is confirmed, not by analysing the art Guiness produces, but by his demonstrating other symptoms like suddenly standing motionlessness with a blank expression and that he self-reports experiencing dizziness, lack of motor control and headaches. A subsequent MRI scan reveals a tumour in the temporal limbic area. In preparation for the surgical removal of

Figure 2.4. Dr. Catherine Black speaking to the parents of the patient painting on a wall and the doors of a hospital room. The painting resembles the work of expressionist artists such as Willem de Kooning, Joan Mitchel, Asger Jorn and Karel Appel. See Black Box, *season 1, episode 1: 'Kiss the Sky'. First broadcast on 24 April 2014.*

the tumour, Black tells Guiness that he will become his old self again as a result of the procedure. Guiness responds by asking if he will still be able to draw after the surgery, to which Black answers: '[Of] course you will! You may not want to anymore'. In the final segment of the show pertaining to this case, it is indeed revealed that the patient has lost his urge to paint after undergoing surgery. Further, looking at the wall paintings in the hospital room, he expresses disbelief that he actually made them. While this case problematises the idea that an accurate diagnosis can be achieved through analysing art, it also reinforces the idea that neurological afflictions and creativity are closely connected.

A similar case, where a person's artistic drive and ability are framed as stemming from a malfunction in the brain, is presented in the documentary *Brain Story*. This non-fiction example further underscores our argument that neurological disease is commonly being used as an explanatory factor for creative abilities. The host, Susan Greenfield, introduces Dick Lingham in a voiceover. He has been diagnosed with a, not further specified, degenerative brain disease that is 'slowly destroying the front of his brain'. Concurrent with the degeneration of Lingham's frontal lobe, Greenfield explains that 'the brain damage has released abilities that Dick never knew he had; he has become over-whelmed with the urge to paint'.

In the segment, Lingham is interviewed while he paints, and the pro-gramme visually alternates between showing his face and a close-up of the paper in front of him. His artworks, Lingham explains, are made by building up a pattern using diluted ink which he later tries to 'inter-pret into some reasonable picture'. The programme then cuts to one of his paintings, which seems to be depicting a dragon and a rearing horse with a rider on its back, predominantly painted in shades of blue (see figure 2.5).

In the interview, Lingham says that he enjoyed doing art at school but that he was not very skilled at painting when he was younger and that as an adult he had not painted until he got sick. His return to painting as a result of a neurological affliction is reminiscent of the artistic urge seen in Guiness in *Black Box*. Stylistically, Lingham's work also bears a resemblance to the wall paintings created by Guiness; the work of both artists brings to mind the painterly style associated with Expressionism. As mentioned previously, this modernist art movement is strongly con-nected to the idea of art as a pure expression, where the artist's mind transcends the body and is projected onto the medium.

In both the real-life case of Lingham and the fictional case of Guiness, altered neurological processes due to brain disease are used as the sole explanation for these patients' sudden urge and ability to paint. However, studying human behaviour, actions and personality traits from a neuro-logical perspective alone leaves out aspects that are crucial for under-standing the interaction between the individual, language, context and culture. From a wider, socially oriented perspective it is highly probable that major mental or cognitive transformations will affect a person's per-ceptions of his or her body and self. Suffering from a disease that changes one's way of interacting with other people, one's ability to work and care for one's loved ones, or that alters one's sense of self, will most likely affect the need to express oneself in order to deal with the situation. In other words, it cannot be ruled out that the connections found between neuro-logical changes and new ways of expression is to some extent the result of social causes. Thus, the criticism of an over-simplified notion of the divide between the individual and the surrounding society that has been directed at modernism's conception of the artistic genius since the lat-ter half of the twentieth century can also be applied to popular media's

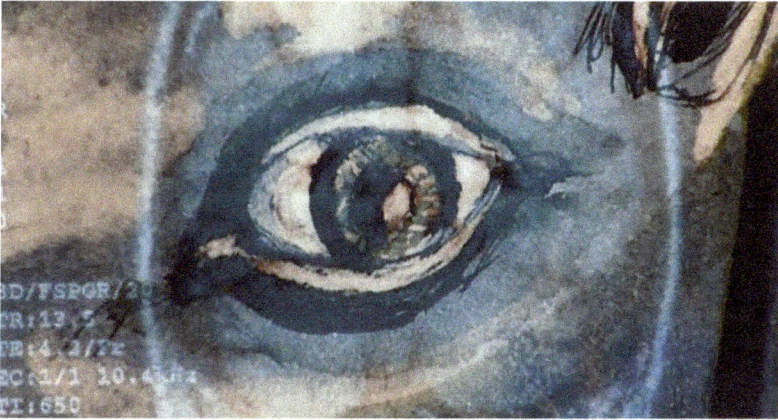

Figure 2.5. A close up of a detail of one of Dick Lingham's paintings. In the documentary Brain Story, *the painting is merged with a brain scan image through a crossfade. This emphasises visually the connection between Lingham's neurological affliction and his art. See* Brain Story, *episode 1: 'All in The Mind'. First broadcast on 8 May 2004.*

contemporary narratives of subjectivity. By furthering the idea of society as consisting of isolated individuals, these framings ignore the complex interaction of bodies, perceptions, language, context and social experiences, which is part of the basis for the formation of subjectivities and personality traits.

The idea of the sole artistic genius may nevertheless be a convenient, even necessary, trope to draw on in popular media when relating creativity to neurological disease. While it to some extent misrepresents what can be inferred from artworks, this familiar shorthand serves to draw the viewer into the narrative in virtue of its recognisability. Acknowledging that art is actually the product of a larger system would muddle the understanding of the artistic vision as having its nexus in mental illness. Popular media programmes dealing with neuroscience repeatedly depict artists as having a certain kind of brain geared towards creativity, and are hence reproducing the modernist idea of subjectivity. The main difference is that the previous focus on the special mind or sensibility of the artist has been replaced by a focus on the brain. By using these specific frames of reference, popular media essentially stage artistic ability as a symptom of an affliction in the organic brain and explain creativity solely through neurological processes. This description of individual expression portrays the biological body as an isolated system and over-looks the importance of other factors (for example social and psychological) that may influence an artist's creative work.

Conclusion: framing neuroscience in popular media

Popular media products can be useful vehicles for framing and relating information about complex topics and creating debate among a broader public. Close readings of popular media products and their narratives can potentially help understand aspects of discursive transformations that are presently underway within science and society at large. As a case in point, we have looked at how the popular science-programme *Brain Story* and the fiction series *Black Box* frame and present neuroscientific themes and findings in an accessible manner. Viewing these and similar media products can thus be seen as a step towards enabling and encouraging an audience to engage actively in a discussion about neuroscience and neurological disease. However, it is important to note that exposure to programmes related to a specific theme does not necessarily lead to such active participation. While popular media can certainly serve to encourage participation, the latter only really occurs when audiences engage critically with the content.

When media products frame complex issues, they often do so by relying on simplification and well-established tropes. The way a given issue is framed will, in turn, impact how an audience may understand and engage with that issue. An example of this is the linking of creativity to neurological affliction, which we have found both in popular science programmes and fictional storytelling. Although it certainly cannot be ruled out that neurological disease can influence a person's creative output, the framings and narratives that are present in our empirical material largely overlook the importance of factors like social environment and practical circumstances. From watching programmes like *Brain Story* and *Black Box*, then, viewers may be left with an affirmation of the old trope that creativity and mental illness go hand in hand and that artistic expressions can be understood as symptoms of neurological afflictions. The truth, like the brain itself, is infinitely more complex.

References

Beaulieu, Anne 2000: The Brain at the End of the Rainbow: The Promise of Brain Scans in the Research Field and in the Media. In: Janine Marchessault & Kim Sawchuk (eds.) *Wild Science: Reading Feminism, Medicine and the Media.* London: Routledge.

Beaulieu, Anne 2002: Images are Not the (Only) Truth: Brain Mapping, Visual Knowledge and Iconoclasm. *Science, Technology and Human Values*, 27(1): 53–86.

Becker, Howard S. 1982: *Art Worlds.* Berkeley, Los Angeles and London: University of California Press.

Bourdieu, Pierre 1993: *The Field of Cultural Production: Essays on Art and Literature.* New York: Columbia University Press.

Demarin, Vida 2009: Neurological Disorders in Famous Painters. *Journal of The Neurological Sciences*, 01/2009.

Dumit, Joseph 1997: A Digital Image of the Category of the Person. PET Scanning and Objective Self-fashioning. In: Gary Lee Downey & Joseph Dumit (eds.) *Cyborgs and Citadels: Anthropological Interventions in Emerging Sciences.* Sante Fe: School of American Research Press.

Dumit, Joseph 2004: *Picturing Personhood: Brain Scans and Diagnostic Identity.* Princeton: Princeton University Press.

Fonsmark, Anne-Birgitte 2014: *Van Gogh, Gauguin, Bernard: Dramaet i Arles.* Charlottenlund: Ordrupgaard/Strandberg Publishing.

Jenkins, Henry 2006: *Convergence Culture: Where Old and New Media Collide.* New York and London: New York University Press.

Kester, Grant H. 2011: *The One and the Man:. Contemporary Collaborative Art in a Global Context.* Durham: Duke University Press.

Nisbet, Matthew & Declan Fahy 2013: Bioethics in Popular Science: Evaluating the Media Impact of The Immortal Life of Henrietta Lacks on the Biobank Debate. *BMC Medical Ethics*, 14: 10.

Nisbet, Matthew & Ezra M. Markowitz 2014: Understanding Public Opinion in Debates over Biomedical Research: Looking beyond Political Partisanship to Focus on Beliefs about Science and Society. *PLoS ONE*, 9(2).

Pąchalska, Maria, Leszek Buliński, Bożydar L.J. Kaczmarek, Bożena Grochmal-Bach, Grażyna Jastrzębowska & Maria Bazan 2013: Fine Art and the Quality of Life of a Prominent Artist with Frontotemporal Dementia. *Acta Neuropsychologica*, 11(4): 451–471.

Piechowski-Jozwiak, Bartlomiej & Julien Bogousslavsky 2013: Neurological Diseases in Famous Painters. *Progress in Brain Research*, 203: 255–275.

Porter, Roy 2009 [1999]: Medicine. In: Iain McCalman, Jon Mee, Gillian Russell, Clara Tuite, Kate Fullagar & Patsy Hardy (eds.) *An Oxford Companion to the Romantic Age.* Oxford: Oxford University Press.

Rose, Nikolas & Joelle M. Abi-Rached 2013: *Neuro: The New Brain Sciences and the Management of the Mind.* Princeton: Princeton University Press.

Uitert, Evert van 2016: Gogh, Vincent van. In: *Grove Art Online: Oxford Art Online.* Oxford University Press, http://www.oxfordartonline.com.ludwig.lub.lu.se/subscriber/article/grove/art/T033020 (accessed 24 February 2016).

Whitney, Charles A. 1986: The Skies of Vincent van Gogh. *Art History*, 9(3): 351–362.

Zaidel, Dahlia W. 2014: Creativity, Brain, and Art: Biological and Neurological Considerations. *Frontiers in Human Neuroscience*, 8: S389.

3. 'Biospace': Metaphors of space in microbiological images

MAX LILJEFORS

'Space' is a polysemous word. It can refer to physical space understood as an objectively measurable container of objects, and it can mean the perceptual space that we experience subjectively through our senses. Even between these two understandings of space (there are of course numerous others) a contradiction appears. We think of physical space as an empty repository that pre-exists any object that is placed within it. Perceptual space, on the contrary, emerges to a large extent *from* objects, like a force field or radius of influence emanating from them. For instance, when we look at the sky we know that physical space extends forever above us, but we tend to perceive 'the sky' like a ceiling located at a height three or four times above the tallest object around us (Arnheim 1978: 25–26). When decorating, we want things to have the 'proper' distance to one another and arrange them accordingly. To make an aesthetic composition is to shape perceptual space by arranging things in physical space.

I will show in this essay, that perceptual space plays an important role in how microscopic images, or micrographs, make us conceive of the interior of the body. Micrographs do not only allow us to behold the inside of our bodies on a minuscule scale but they also *compose* this diminutive dimension into something we can comprehend and share, i.e. into a body of knowledge and meaning. Therefore, I will argue, to understand the cultural meanings of micrographs it is not sufficient to examine only how cells, proteins, and other biological elements, are made to look in micrographs. We must also ask *where* micrographs represent them as existing.

Many philosophers have explored how physical space and places are organized into social, political and discursive sites and spatialities; the

49

works of Henri Lefebvre (1991 [1974]), Gaston Bachelard (1994 [1958]), Edward Casey (1997), and Miwon Kwon (2002) are seminal. The very concept of space harbours a promise of order, a possibility to map positions and distances, to orient oneself. Therefore, spatial metaphors are very often used to make abstract ideas easier to grasp, although such metaphors may distort their nature. In *Civilization and Its Discontents* (2010 [1930]), Freud lamented the fact that psychoanalysis had to rely on spatial metaphors to explain the dynamics of psychic processes. To bring home the insufficiency of these metaphors, he encouraged his readers to attempt the opposite intellectual operation: to imagine an actual place, such as the city of Rome, as if it had the features of a psyche. Just like different stages of psychic development may manifest themselves simultaneously in a person's mental life, Freud explains, Rome's different epochs would coexist in the same place at the same time. For instance, the Renaissance Palazzo Caffarelli and the Roman Tempel of Jupiter Capitolinus, on the ruins of which the Caffarelli palace was erected, would occupy the same location simultaneously, one emerging through the other by a mere shift of the beholder's angle of observation (Freud 2010 [1930]: 16–18). Freud's absurd image is intended to illustrate the inadequacy of spatial metaphors for representing psychic dynamics, but thereby it also brings home the degree to which spatial metaphors confer a sense of orderly logic to the phenomena they are made to represent. We may add to this, that Freud's thought experiment also tells us something important about how pictures express meaning. In analyses of pictures, like the ones in this essay, images are often described as having 'layers' of meaning; we talk about 'surface' value, about probing 'deeper' and 'uncovering', and so forth. These common figures of speech make up a spatial metaphorics, which on the one hand gives a sense of order to the analysis, but on the other belies the fact that meanings in images do not exist 'separated' from each other, and that it is the interpreter's angle of observation that determines what meanings emerge into view. Of course, by talking about 'angles' and 'emerging' we employ still more spatial metaphors to distinguish various modes of regarding things.

I shall begin this essay with a picture that may at first seem remote from micrographs: the first high-resolution colour television image of planet Earth, taken by the American satellite ATS-3 on 10 November 1967 (see figure 3.1). What does this oft-reproduced picture, made up of 2 400 horizontal lines each representing a width of almost four kilometres, tell

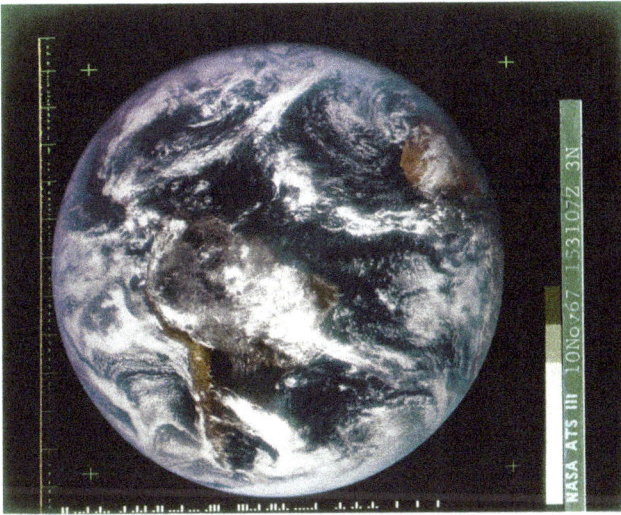

Figure 3.1. Earth seen from space. The picture was taken by the NASA satellite ATS-3 (Applications Technology Satellite 3), on 10 November 1967. Published in: Edgar M. Cortright, (ed.) (1968) Exploring Space with a Camera. *Washington D.C.: National Aeronautics and Space Administration, p. IV. Image source: Wikimedia Commons.*

us about how microscopy portrays the diminutive dimensions of our biology? As I shall discuss in the next section, biological entities are often depicted as if they exist in a vast and empty cosmic space rather than inside actual bodies. Therefore, this iconic image of Earth gives us a hint of the visually mediated meanings that accompany similar representations of space. The picture was published by NASA in 1968, in the photo book *Exploring Space with a Camera* (Cortright 1968). By that time, space photographs of Earth had existed for several years but had not been released to the public. NASA's book was preceded and perhaps partly occasioned by a grass-root campaign with the rhetorical slogan, 'Why haven't we seen a photograph of the whole Earth yet?', which seems to indicate that a sense of the impact such an image would have was already in the air (Nisbet 2014: 71). In any case, once published, the image of the Blue Marble, as this motif came to be named, inspired Marshall McLuhan's notion of 'the global village' as well as James Lovelock's Gaia thesis, and has continued to inspire various forms of planetary awareness, the thought that humanity shares one world. It should be noted, however, that the Earth image initially also spurred other, less positive associations (see Nisbet 2014: 81–86), but over time the positive ones have prevailed. Some commentators have taken a more critical stance. Spivak (1998: 329–330) remarks about McLuhan's concept, that nobody actually lives in the *global* village because Earth appears flat, not as a globe, to all its inhabitants. It can be envisioned as a spherical object only from a cosmic distance. From such a position, all kinds of uneven power relations disappear from view just like the topography of any particular terrain. The Earth as a globe is accessible only to the sensory modality of sight. It is marked by a remoteness that is the hallmark of universalizing ways of looking and thinking, Spivak points out. Nisbet, too, sees that the image of the globe implies simplified solutions to complex problems, as conflicts and tensions disappear behind an emblematic symbol of humanity as one undivided family (2014: 80–86).

From an art historical point of view, the Blue Marble image can be said to constitute the final step of the tradition of landscape painting. Several art historians have discussed how looking out over a scenery from a high vantage point – in a painting or in real life – conveys the impression of 'a world', a totality larger than the human imperfections and conflicts that are encompassed within it. Viewing a landscape-as-world instils a calm and passive attitude in the beholder. Alois Riegl wrote about his view from an Alpine peak, 'nothing is near or within reach, and nothing stirs my sense of touch. Looking is everything here' (1929 [1899]: 28). Verschaffel considers that distance is the landscape's crucial trait: 'the landscape, an independent world that is no longer an environment, appears in front of the beholder – not here, but there' (2012: 2). He goes on to explain that the landscape is thus an image and it shares its remoteness with all images, regardless of their subject matter, because every image opens up a space before the beholder, which she can look into but not enter or act upon. Only the landscape picture realizes the inherent stillness of every image, Verschaffel (2012: 2–3) concludes. Nonetheless, the Blue Marble motif takes that detachment one step further, by getting rid of the horizon. In the landscape, the horizon line, whether visible or implied, establishes a connection between picture and viewer, because it indicates the height and distance of the observer's point-of-view. In the motif of the Earth, however, the horizon has turned into the contour of the spherical planet itself, and, thus biting its own tail, it unties any specific relation to the beholder. Symmetrically framed in the black square, the illuminated globe rests in perfect self-containment. Cosmic space and pictorial space coalesce into an image of Earth as 'world', distant and complete. Below, I shall show that this remoteness and detachment of the cosmic image-landscape is essential also to the aesthetics of biological micrographs, in spite of their vast difference in scale.

Microscopic biology is made to resemble cosmic space

Figure 3.2 presents a montage with nine microbiological pictures, and figure 3.3, one with nine astronomical pictures. A glance suffices to show how much the two kinds of images resemble each other with their luminous, multi-coloured shapes hovering weightlessly against a black, empty background. Biological entities are portrayed as celestial bodies or strange deep-sea creatures glowing with a light of their own in a realm of otherworldly majestic quietude. It is this 'other world' that is my topic, the perceptual space that is produced around the biological entities depicted in micrographs. I shall call this space, for short, *biospace*. It is purely pictorial, which is to say, that it emerges and exists only in and through pictures. As in the Blue Marble image, biospace sets up a relation of simultaneous detachment and visual fascination between beholder and object. It also functions, I shall argue, as a shared cognitive space, a 'realm of understanding', in which biological life can be enveloped in socio-cultural meanings.

To imagine that the external world is built from the interior of a body is an old custom. Several ancient creation myths tell how the world was built from the corpse of a slain archaic monster. In Babylonian mythology, the god Marduk creates heaven and earth from the carcass of the dragon Tiamat. In Nordic mythology, the brother gods Odin, Vile and Ve slay the frost giant Ymer and build the world from his severed body. The poem 'Grímnismál' in the *Poetic Edda* (2004: 37) tells how the gods created the sea from Ymer's blood, the hills from his bones, the sky from his skull, and from his brain, they shaped the clouds. These myths suggest, that the world that humans inhabit has risen from the remains of a giant *ur*-creature.

The idea of a micro-macro correspondence between humans and the universe runs through Western philosophy, from Antiquity throughout much of the Middle Ages, and is revived in early anatomical science during the Renaissance. As scholars Barbara Stafford (1993) and Jonathan

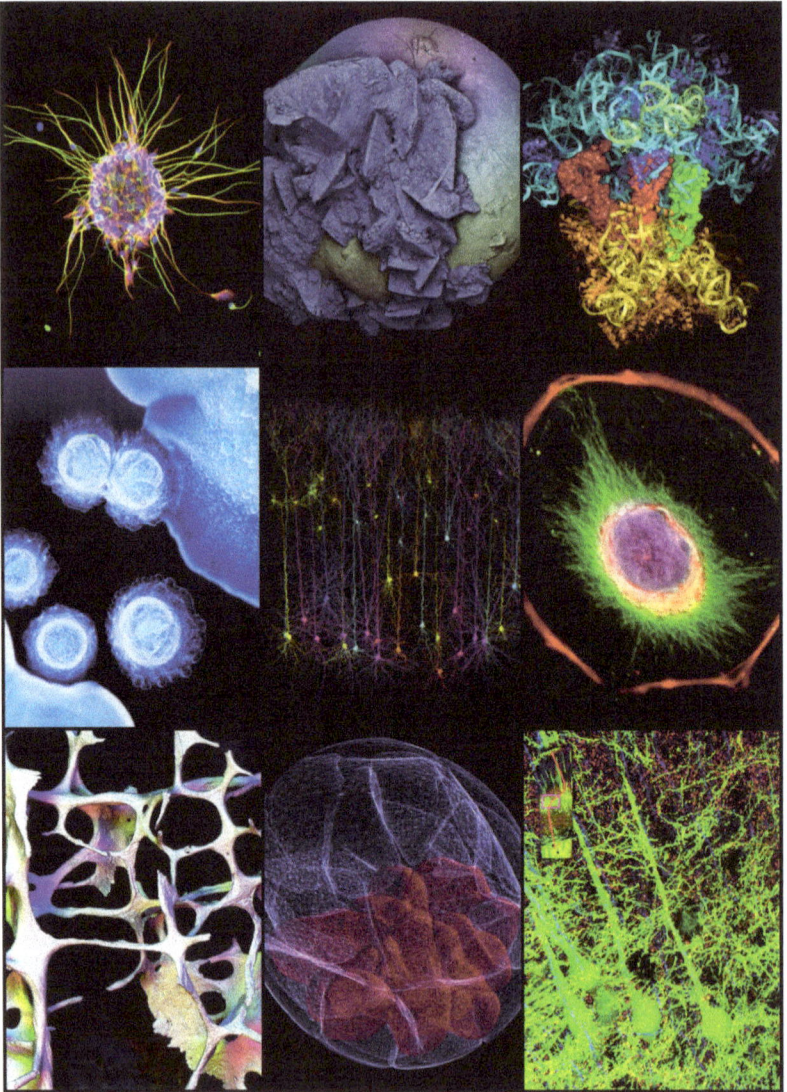

Figure 3.2. Montage of nine bioscientific images, slightly cropped. From left to right, starting at the top: fluorescence micrograph of neurons in culture, scanning electron micrograph of a kidney stone, molecular model of a bacterial ribosome, colour-enhanced photomicrograph of periodontal bacteria, computer simulated image of synthetic pyramidal neurons, confocal micrograph of a 3T3 fibroblast cell, scanning electron micrograph of osteoporotic bone, computer 3D reconstruction of a mouse embryo, video still from array tomography 3D reconstruction video Machinery of Mind. *For more details and image sources, see list at the end of this chapter.*

Figure 3.3. Montage of nine astronomic images, slightly cropped. Credit: European Southern Observatory. From left to right, starting at the top: Antennae Galaxies composite of ALMA and Hubble observations, VISTA's infrared view of the Cat's Paw Nebula, the Helix Nebula imaged with the Max-Planck Society/ESO telescope 2.2, the Tinker Bell Triplet by ESO Very Large Telescope/Hubble Space Telescope, Thor's Helmet Nebula imaged with Very Large Telescope, the Dumbbell Nebula by Very Large Telescope, a 340-million pixel starscape from Paranal, Very Large Telescope image of the cometary globule CG4, Very Large Telescope image of the Arches Cluster. For more details and image sources, see list at the end of this chapter.

Sawday (1996) have shown, the rise of empirical anatomy and the practice of dissection led to novel ways of representing the bodily interior as a world of its own. The anatomized body became a primary cognitive model for other composite unities, like architecture, society, or the cosmos. Moreover, geographical metaphors were used to portray dissection as a discovery and exploration of an external territory. The anatomist was likened to an explorer of an unknown continent, who accumulated his knowledge in an anatomical *atlas* – a term coined by the founder of modern cartography, Gerardus Mercator – where the bodily interior was mapped with names and demarcation lines. To some extent, the two sciences developed side by side: Andreas Vesalius, whose seminal seven-volume opus *De Humani Corporis Fabrica* (from 1543) inaugurated empirical anatomy, studied in Leuven at the same time as Mercator. Like the Columbian explorers before him, Vesalius claimed that first-hand observation and ocular evidence, not hearsay or tradition, was the methodological foundation of his findings. Therefore, when the biosciences today depict our biological interiority as a cosmic space, they do not create an entirely new spatial metaphor but a variant of an already established tradition of metaphors. What is striking with this new variant is that the diminutive scale of microscopy seems to inspire the grandest metaphors. Anatomy compared the dissected body to earthly terrains, whereas the biosciences today mobilize the infinitely larger expanse of the cosmos – the more *micro* the object, the more *macro* the metaphor.

Biological entities are often presented in micrographs against a black background that gives the impression of emptiness behind them. Where does this background originate from? One can say, that it is produced *ex nihilo*, from nothing. Black signifies the absence of data; a sufficiently strong signal may have been lacking in the measuring apparatus, or a registered signal may have been masked off in the visualizing process. In any case, black usually denotes 'no data', a meaning that marries easily with the visual association of empty depth. Absence of data, however, has no natural colour, obviously. Black is part of a colour code and could, in principle, be substituted with any other colour or pattern. 'So why, then, is black used so consistently?' I once asked a professor of radiology. Her reply combines function and aesthetics: 'Colours look so clear against black. And it is beautiful too.'

Distinct visualization of data as well as pleasing appearance; we may here distinguish between an epistemological and an aesthetic 'layer' in the image. We can also regard it as denotative data and connotative meaning, showing different 'dimensions' in the image. But then, we are employing idealized concepts to stratify analytically that which in reality meets the eye of the beholder as one unified visual expression. The latter is true particularly if the beholder is a lay person, who knows little about scientific imaging technologies. It is therefore important to note that the microscopic images in figure 3.2 as well as the astronomic images in figure 3.3 have been produced for the purpose of public dissemination. The micrographs – except the one to the lower right, more about that below – have all been rewarded with the Wellcome Image Award, an annual prize from the Wellcome Trust, a major British financier of medical research. They are accessible at no cost from their webpage. The astronomic images belong to the 'Top 100 images' of the European Southern Observatory (ESO), and are also free to download. This indicates that scientific images do not trickle passively into society at large. Scientists intentionally make their images aesthetically gratifying and actively spread them to lay audiences. How, then, does the visually pleasing appearance of much of the scientific imagery influence how lay audiences understand, or misunderstand, science?

Art historian Annamaria Carusi (2008) has proposed the hypothesis that the aesthetic features of scientific images can contribute to trust and community building among scientists. Drawing on Kant's philosophy of aesthetics, Carusi suggests that conventions for how images 'should' look are part of a shared framework of understanding among scientists, a *sensus communis,* which, in turn, is a prerequisite for collaboration between specialists with differing expertise in large-scale science programmes (p. 253). Here, Carusi distinguishes between *in-silico* visualizations, i.e., computer simulations used for instance in computational biology, and images that represent actually existing objects, such as micrographs (p. 246). I believe that this distinction has significance in scientific communities. But it is reasonable to assume that it is less important for lay persons, who lack specialized knowledge about microbiology as well as about differences between specific visualization techniques. Nevertheless, I suggest that the homogenous appearance of microbiological pictures of various kinds can contribute to a shared understanding of microscopic biology among the general public. The reason for this assumption is twofold. Firstly, Kant's theory of aesthetic judgement and *sensus communis,* in *Critique of Judgment* (1952 [1790]), has universal scope and is not restricted to particular professional interests. Secondly, because the scale of microbiology is inaccessible to unaided human sight, lay audiences cannot determine whether micrographs of cells, molecules, and other entities, are visually accurate or not. Therefore, as I have argued elsewhere (Liljefors 2012), it makes sense to view the visual culture of microbiology as a repertoire of aesthetic tropes, which contributes to a collective imagination about the microscopic dimension of our bodies. In the trail of recent progresses in biomedicine, that dimension is rapidly becoming exposed to various economic, political and juridical powers. While I can only briefly touch on that issue here (in the first of my three conclusions below), I think it is striking that human biology is visualized as a remote, otherworldly realm at the same time as it is turned into a real economical asset in this world. In the next two sections I shall examine in more detail how that aesthetic convention is disseminated through culture at large.

Biomedical images are distributed to lay audiences

Images have become more frequent in science communication. This fact is illustrated in an infographic made by media theorist Lev Manovich and his colleagues at Software Studies Initiative (see figure 3.4). The picture is simultaneously a photographic montage and a diagram. It is composed of scanned pages from the magazine *Popular Science*, 1882–2007, one issue from every fifth year laid out chronologically starting from the upper left. In that respect, we are looking at a photograph of 150 years of popular science. At the present resolution, we cannot see the content of the pages but they become noticeably darker in the lower part of the infographic. The reason for this is that over time pictures have come to occupy an ever-larger portion of the page area and, proportionally, text takes up a smaller part of the page. During its first decades, *Popular Science* included only occasional pictures. It was less oriented towards publishing pictures than the scientist-oriented journal *Science*. Today, that situation is reversed. Manovich's montage can thus also be read as a diagram, in which the gradual darkening of the magazine pages represents a quantitative increase of the number and size of images used in science communication. From having functioned earlier as a supplement to written accounts of science, images have increasingly taken centre stage. Our expectations have grown on the capacity of images to communicate scientific knowledge to the public.

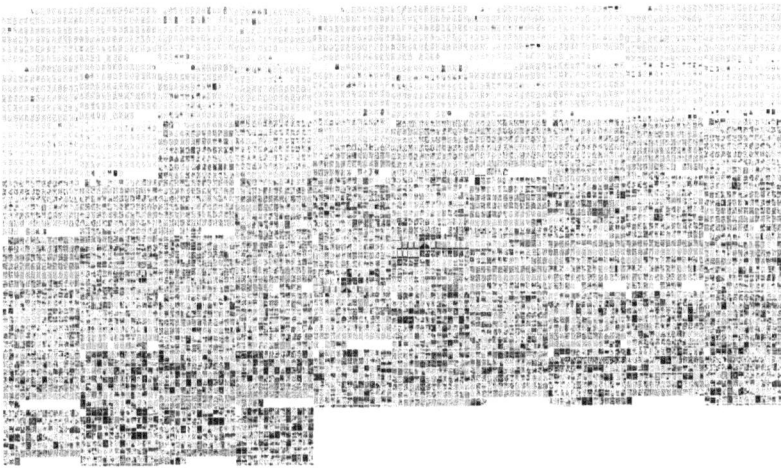

Figure 3.4. Visualization of image use in Popular Science *magazine, one issue per every five years, 1882 to 2007. Credit: William Huber, Tara Zepel and Lev Manovich, Software Studies Initiative 2010, available from http://lab.softwarestudies.com/2010/11/science-and-popular-science-magazines.html.*

Another indication that aesthetics is deemed important in science communication is the numerous prizes that have been created for scientific pictures. The above-mentioned Wellcome Image Award started in 1997. Since then, numerous more image awards have been instated by medical research institutions, science foundations, and manufacturers of imaging instruments. The assessment criteria of these image awards often reveal that aesthetic attributes are central. For instance, the International Science & Engineering Visualization Challenge, includes the following criterion:

Visual Impact: A successful entry provides viewers with new scientific insight, is visually striking, and conveys the artist's skill and expertise in the chosen medium (e.g. photography). The entry also conveys the artist's mastery of the seven fundamentals (color, value, line, texture, shape, form and space), and principles (balance, emphasis, harmony proportion, variety, gradation, movement and rhythm) of design.[1]

The Lennart Nilsson Award is given to images that '[reveal] science to the world in beautiful, unique and powerful ways',[2] while The Koch

1. International Science & Engineering Visualization Challenge. Retrieved from: http://www.nsf.gov/news/special_reports/scivis/judging.jsp (accessed 28 February 2017).
2. The Lennart Nilsson Award. Retrieved from: http://www.lennartnilssonaward.se/the-foundation/ (accessed 28 February 2017).

Institute Image Awards promotes 'extraordinary visuals' that are 'beautiful and thought-provoking'.[3] The Drosophila Awards encourages 'compelling images',[4] The Art of Science Image Contest aims to 'showcase the artistic side of scientific imaging', based on 'artistic and/or visual impact of the images',[5] while the Great British Bioscience Image Competition rewards pictures that show 'the beauty of science and its impact'.[6] This quote from the Nikon Small World competition sums up the approach of scientific image contests:

> A photomicrograph is a technical document that can be of great significance to science or industry. But a good photomicrograph is also an image whose structure, color, composition, and content is an object of beauty, open to several levels of comprehension and appreciation.[7]

The prizes indicate that aesthetic pleasantness is not some insignificant side-effect but a central pillar in visual communication of science. Awarded images are exhibited at science museums, covered in magazines and newspapers, and highlighted on webpages and YouTube channels. Winning entries in the GE Cell Imaging Competition are displayed on a large screen on Times Square, New York. It is important, however, not to exaggerate the divide between the contexts of the laboratory and the public sphere. An image is never only a representation of data to experts, nor only an aesthetic image to lay people. During my field studies in radiology departments, I have observed radiologists admire the aesthetic qualities of an X-ray or MRI picture at the same time as they were assessing the medical information it represented. Most lay viewers are presumably aware that they lack the knowledge to decipher all the information that is contained in a scientific image. That said, images are usually presented quite differently in scientific and in popular contexts. A case in point is the Digital Imaging and Communications in Medicine standard,

---

3. The Koch Institute Image Awards. Retrieved from: http://ki.mit.edu/approach/imageawards (accessed 28 February 2017).

4. The Drosophila Awards. Retrieved from: http://www.drosophila-images.org/ (accessed 28 February 2017).

5. The Art of Science Image Contest. Retrieved from: http://www.biophysics.org/Awards/SocietyContests/ArtofScienceImageContest/tabid/4124/Default.aspx (accessed 28 February 2017).

6. Great British Bioscience Image Competition. Retrieved from: http://bbsrc2014.picturk.com/competitions (accessed 28 February 2017).

7. Nikon Small World competition. Retrieved from: http://www.nikonsmallworld.com/photo/info (accessed 28 February 2017).

or DICOM for short. It is an international standard for organizing and processing information in medical images, with the purpose of ensuring adequate communication between technological platforms and between experts. Put simply, a DICOM file consists of an image and a protocol for metadata, such as information about the patient, the imaging procedure, pixel data, etc. The metadata is necessary for researchers and clinicians to make diagnoses and take decisions about further experiments or treatments on the basis of the image. DICOM includes fields for thousands of different kinds of metadata. Only a portion of them are used for any given image, but the number hints at the extent to which scientific images are dependent on non-visual information to be functional in scientific contexts.

Metadata, however, rarely accompanies a scientific image when it is disseminated to lay audiences. Typically, a caption will provide basic information about what the image depicts and perhaps about the technology that was used to acquire it. In most cases, little is explained about the technical, chemical, and algorithmic details of the imaging process. Instead, the image stands 'on its own', conveying meaning through its visual appearance alone. Without keys for decoding the data that the image originally was intended to express, the image appears un-coded and epistemologically transparent. In the same process, it becomes available for projections of more existential meanings.

A descent into the brain-mind

Now follows an example of how differently a microscopic image can be presented in scientific contra public discourses. In figure 3.2, the image in the lower right is a still frame from a video produced by the Smithlab neuroscience laboratory at Stanford University. Smithlab has its own YouTube channel where researchers make videos available to the public. The image reproduced in figure 3.2 is from Smithlab's most-viewed video, *Machinery of Mind* (2012), in which the viewer is taken on a simulated flight through the cortex of a mouse brain. The mouse cortex has been visually reconstructed in 3D through an imaging technology called array tomography. Array tomography was invented at Smithlab and allows high-resolution volumetric imaging of tiny biological ultrastructures such as the molecular architecture of brain tissue. The technique was presented to the scientific community in two articles in the journal *Neuron* (Micheva & Smith 2007; Micheva et al. 2010), and to the public through the YouTube video *Machinery of Mind*. I shall here first attempt to summarize the procedure of array tomography in a manner accessible to lay readers. Being an art historian, I am a lay person myself in this field, and the point of my description is less to communicate science than to make explicit the limitations of a lay person's technological understanding. Beyond that limit, there is a domain of deeply specialized expertise, which I shall go on to exemplify with a few quotes from the *Neuron* articles. Finally, I shall make an aesthetic analysis of the *Machinery of Mind* video. The purpose of the comparison is to give a sense of how differently a scientific image is framed in expert versus popular discourses.

Here follows my lay person explanation. Array tomography allows researchers to study extremely thin slices of tissue, about 70 nm (nanometres) in thickness. In comparison, the average diameter of a human hair is almost 1 500 times thicker, 100 000 nm. The array tomography process can be divided into six steps:

1) The specimen (in the video, a slice from a mouse brain) is dissected and embedded in a certain kind of resin.

2) Small blocks of tissue are mounted and sliced in a ultramicrotome, a machine equipped with diamond knives used to cut very thin sections.

3) An 'array', or ribbon, of successive sections is attached onto a glass coverslip. A ribbon is about 45 mm long and can consist of more than 100 sections. The sections stay glued together, edge to edge, because an adhesive has been applied to the top and bottom of the tissue block before it is sliced.

4) The array is then stained with a method called immunofluorescence. Staining means that a dye is injected into the specimen, which marks or 'tags' a specific component in it, making it stand out in the image. In immunofluorescence, a so-called antigen (short for 'antibody generator') is injected. An antigen activates antibodies, i.e. an immune response, in an organism. Different antigens attract specific antibodies, which allows researchers to target selected components in the specimen. The antigen has been attached with a fluorophore, a chemical compound that emits fluorescence when exposed to light,

which makes the antibody, to which the antigen bonds, visually detectable.

5) After the staining, the array of sections is imaged with a microscope. (Different kinds of microscopes can be used.) In this process, the specimen is exposed to light, which triggers the fluorescence of the fluorophore, which in turn is registered by the microscope. Every section in the array is imaged separately, which results in a sequence of images representing the entire array.

6) In a computer, the images are aligned into a stack. The image stack will constitute a three-dimensional simulation of the original tissue block, because the images are placed in the same order as the sections that were sliced from the block. In this 3D reconstruction, the spatial relations between neuronal structures are represented. When the simulated 'camera' flies through the imaged brain section, the strands of neuronal tissue move realistically in relation to each other.

To sum up, a tissue block is sliced into a series of cross sections. The sections are lined up and imaged in a microscope individually. The resultant series of images is stacked into a digital 3D 'image block', which represents the original block of tissue.

This is my best attempt as a lay person to explain array tomography to other lay persons. It is, however, a very simplified account, a fact that I shall highlight with a few random quotes from the *Neuron* articles, of what I have left out. Specifications of the laboratory mice used in the research: 'C57BL/6J', 'YFP-H', and 'Sprague Dawley rats' (Micheva & Smith 2007: 34). The dosage of immunofluorescence staining: '50 mM glycine in Tris buffer (pH 7.6) for 5 min' (p. 35). Details of the eluting process between repeated staining: 'dH2O and a solution of 0.15M KMNO4 and 0.01N H2SO4 was applied for 90 s.' (p. 35) Those brief examples should suffice to indicate that beyond my simplified account, there is a world of specialized knowledge that is required for actually using array tomography.

In the *Machinery of Mind* video, we find an altogether different form of representation, which does not rely on a simplified textual description but on a dense audio-visual rhetoric. The video begins with a photograph of a (presumably) living mouse in green grass (see figure 3.5, upper left). An image of the mouse brain fades in and detaches itself, while the mouse and the grass fade out into black. The transparent brain hovers alone against a black background (see figure 3.5, upper right). A small part of the brain begins to glow green. The simulated camera zooms in on that section and starts a slow descent into the mouse brain (see figure 3.5, lower left and right). The descent takes the viewer into what I have referred to as *biospace*, an undefined and black emptiness here traversed by luminescent, seaweed-resembling neural tissue. The otherworldly character is enhanced by the camera's smooth movement without any indication of gravity or friction. No light is reflected back from any surrounding surface, which would have implied an enclosed space. The cortex network seems to extend into endless depth on all sides. Nothing indicates that the imagery is created with a technology for imaging extremely *thin* slices of tissue. The video contains almost no technological information, besides the fact that it depicts a mouse cortex and the names of some layers of tissue (see figure 3.5, lower left and right). There is no voice-over, but the entire video is set to spherical music, commissioned by Smithlab from songwriter Catherine Rose Smith. The audiovisual aesthetics, together with the near complete absence of didactic information, result in a poetic rather than an epistemological statement; the video conveys an atmosphere rather than empirical facts. Instead of explaining the technical, chemical and algorithmic complexity of the imaging process, it portrays brain imaging as an odyssey into an inner limitless space – 'inner', not only in a physiological sense, because the word *mind* in the video title, instead of *brain*, implies that we are gazing into a mental cosmos.

Conclusion: what is absent in biospace?

Researchers at the European Southern Observatory (ESO, see figure 3.3) have written an article titled, 'What Determines the Aesthetic Appeal of Astronomical Images?' It outlines six parameters for achieving 'optimal aesthetic appeal' in astronomic pictures, so that they may 'inspire awe, wonder and enthusiasm, and portray the Universe as a fascinating place worthy of exploration' (Lindberg Christensen, Hainaut & Pierce-Price 2014: 20). I cannot discuss in detail the parameters listed, but the general idea put forth by the authors is that scientists (at least astronomers) may modify the appearance of their images, such as their colour and composition, in order to make them more aesthetically pleasing. At a European Science Foundation conference, I had the opportunity to ask ESO

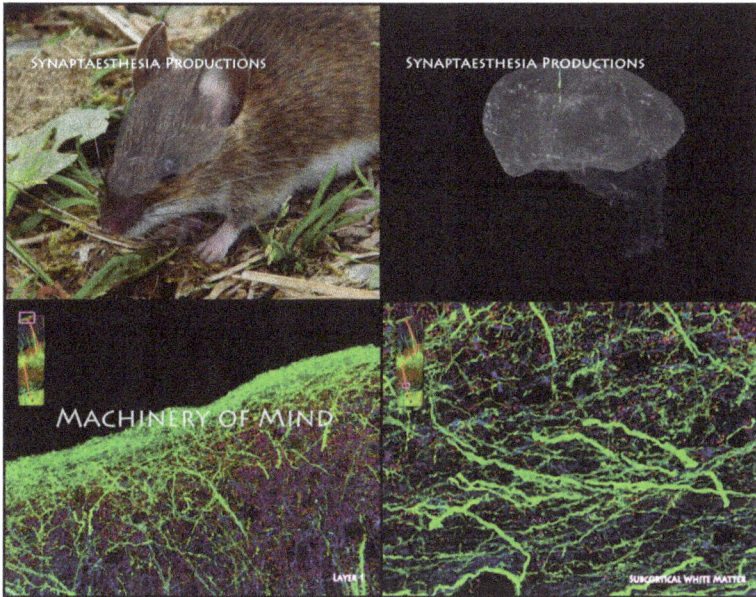

Figure 3.5. Montage of four video stills from array tomography 3D reconstruction video, Machinery of Mind. Credit: Smithlab, Stanford University, available at https://www. youtube.com/watch?v=NFaKLNoItcA.

astronomer Lars Lindberg Christensen, one of the authors of the article (at that time not yet published), where he drew the limit for acceptable image manipulation. He replied that one must never add any new object, like a star, to a picture, if the object does not actually exist out there. Formal modifications, such as increasing colour intensity or cropping a picture, he deemed unproblematic, as I understood him.

This reply made apparent what might be a principal difference in outlook between the natural scientist and the humanist, if such generalizations may apply. For the astronomer, adjusting the colours in a picture is not to 'add' anything and does not change the content of the image. For the art historian, on the contrary, such formal adjustments may very well change the meaning of the picture. The modifications suggested in the ESO article make astronomic images resemble science fiction or New Age imagery to a greater degree, and seem to some extent to be modelled on aesthetic ideals in Romanticist or Expressionist art. Through such connotations, astronomic images tap into a deep reservoir of cultural meanings, foremost perhaps the intuition that nature is permeated by spirit. It is important to note that whatever meanings an image expresses through its aesthetic character, these are rarely put forth straightforwardly as 'statements' about reality. Rather, they are 'atmospheric', like a filter or lens that adds a more or less tangible tinge or mood to the representation. In science, which in one sense is all about making statements about reality, aesthetics can therefore seem to not really matter, to be marginal, and merely decorative. For the same reason, however, the meanings conveyed through aesthetics are immune to refutation. We may find the appearance of any of the pictures in figure 3.2 or figure 3.3 beautiful, or gaudy, but on that ground, it would not make sense to call the picture 'true' or 'false' in a scientific sense. An image may thus express meanings aesthetically that are at odds with assumptions underlying a scientific worldview, or that negate socio-political aspects of the production of the image. I shall end this essay with three brief and tentative interpretations of meanings that are expressed aesthetically in microbiological images. The interpretations have one thing in common: they do not so much address what is seen in biospace as what is absent from it. I ask: what has been omitted from the black and endless emptiness of biospace?

1) *The socio-political context of bioscience is absent from biospace.* In the pictorial imaginary realm that is biospace, biological entities – stem cells, oocytes, neurons, genes, etc. – seem to exist in isolation from the body as a unified organism. Elsewhere, colleagues and I have named this form of dissolved corporeality 'the atomized body', in reference to ancient atomistic theories, which held that reality is composed of isolated atoms surrounded by a void (Liljefors, Lundin & Wiszmeg 2012). In anatomy, the body is divided into 'parts' that usually retain a visually recognizable relation to the body's functional structure. The atomized body, in contrast, is broken down into diminutive 'particles' that seem to exist independently from the body as a whole and from each other. This representational convention mirrors how science nowadays extracts biological particles from bodies and manipulates, reproduces, and studies them in the laboratory (and sometimes inserts them into other bodies). On that level, micrographs reflect the practice of science. But an analysis of a representational paradigm needs also to ask: What is systematically *not* seen in those images? What is absent in biospace? One answer to that question is: the web of economic, political, and legal relations in which bioscience is embedded (see Rose 2007; Lock & Nguyen 2010; Rose & Abi-Rached 2013). The progress of the biosciences has made the body an arena for

commercialization and legislation where interests clash over ownership, copyrights, and the integrity of 'life' and personhood. State power and capitalism meet the hopes and fears of individuals struck by diseases. In other words, while science disentangles our biological particles from their immediate physical context, the body, those particles become increasingly embedded in societal relations of power. This sphere of earthly struggles, however, is absent in the empty blackness of biospace.

2) *Biospace negates the indifference of material reality.* I think there is a kind of secular, theological argument expressed in the visual aesthetics of microbiology. While it would be anathema to the scientific worldview to assume that a divine or higher meaning is the essence of reality, something of that nature is nonetheless conveyed by the aesthetics of biospace. I call it a *theology of beholding.* I have shown above that scientific images intended for public dissemination are designed to inspire fascination for the beauty and elegance of material reality, whether on the cosmic or the microscopic scale. The various means by which this aim is pursued, I claim, add up to a visual discourse that portrays reality as *meant to be beheld by us,* or, in one word, designed. Through pictures, this discourse thus encourages a perception that is diametrically opposed to the idea that reality is governed by indifferent natural laws without purpose or intention. It constitutes, in fact, a non-verbal variant of what is usually referred to as the teleological, or, the 'design' argument, which holds that the complexity and intricacy of the construction of reality indicates an intentional design, and hence, a conscious creator behind it. The attribution of a teleological purpose is a classical argument for the existence of God, employed by Thomas Aquinas, René Descartes, and others. Today, it is used by creationists against Darwin's theory of evolution (or more precisely, the part of the theory that postulates that natural selection governs the evolution of the species). A non-theist, so-called anthropic version of the teleological argument exists as well. It is advanced by, among others, philosopher Thomas Nagel in his much-debated book *Mind and Cosmos* (2012). But we do not need to go into the specific arguments of the teleological position here, because the aesthetic discourse does not precisely *argue* its point. It persuades by presenting an image of reality draped in a certain atmosphere or tint. Biospace is the product of such a visual, wordless, discourse, and at its heart, I argue, lies the attitude that reality offers itself in its beauty for us to behold. It even seems to say that our beholding *is* the teleological purpose of reality's beautifulness.

3) *Biospace mitigates aspects of bioscience that compromise conventional notions of selfhood.* I believe that the aesthetics of biospace diminish the potentially disturbing effects of understanding selfhood as a result of neural, biochemical processes in the brain. Art historian Glenn Harcourt (1987) has made a similar claim about the function of aesthetics in Vesalian anatomy. It is known, that Vesalius borrowed the appearance of antique sculptures when depicting the écorché figures, the flayed bodies, in the *Fabrica*. For instance, the limbless torsos in *Book V* look like they were made from marble rather than soft flesh. Harcourt suggests that this served to alleviate the sight of the contested practice of dissection and its violation of the human figure. As Julia Kristeva observes in her influential essay *Powers of Horror* (1982: 5), 'seen without God and outside of science [the corpse] is the utmost of abjection', a revolting body-thing from which the very idea of an 'I' is expelled. The scientific illustrations of Vesalius protect the beholder from the abject disintegration of the dissected body into flaccid, decaying tissue and fluids.

Bioscience today hardly produces immediately upsetting motifs, but on reflection, its implications for the understanding of somatic selfhood can be just as radical: the autonomous ego explained as an incidental side effect of biochemical processes inside neurons, genes, stem cells, etc. In my opinion, images of biospace counter these implications and bolster a conventional sense of selfhood through a recurring aesthetic trope: our biological particles are rendered as distinct and clearly outlined objects. Although fascinatingly strange-looking, they tend to display coherence of form and unambiguous delimitation and to be easily distinguishable from their surroundings – a result of visualization algorithms being coded to achieve those effects. Portrayed as stable and isolated entities, our biological constituents come across as analogous to conventional models of identity and selfhood. Thereby, in bioimaging as in Vesalian anatomical illustrations, aesthetics serves to mitigate the challenge from the picture's subject-matter to the self-understanding of the beholding subject.

Each of those three meaning formations in the aesthetic discourse of biospace can, of course, be elaborated further. Together they form what might be called a mythological world-view, if not diametrically oppositional to, then at least quite different from a scientific outlook on reality. For that, we may, or may not, appreciate pictures of biospace. Regardless, their aesthetics achieve the double effect of connecting us with and separating us from the phenomena they depict – much like the view of Earth seen from outer space. Their visual appearance constitutes an accessible interface to a domain of science that is impossible to perceive directly and that is complex to grasp, which at the same time shields us from some of its challenging implications.

References

Arnheim, Rudolf 1978: *The Dynamics of Architectural Form*. California: University of California Press.

Bachelard, Gaston 1994 [1958]: *The Poetics of Space*. Boston: Beacon Press.

Carusi, Annamaria 2008: Scientific Visualisations and Aesthetic Grounds for Trust. *Ethics and Information Technology*, 10: 243–254.

Casey, Edward S. 1997: *The Fate of Place: A Philosophical History*. Berkeley: University of California Press.

Cortright, Edgar M. (ed.) 1968: *Exploring Space with a Camera*. Washington D.C.: National Aeronautics and Space Administration.

Freud, Sigmund 2010 [1930]: *Civilization and Its Discontents*. New York: W.W. Norton & Company.

Harcourt, Glenn 1987: Andreas Vesalius and the Anatomy of Antique Sculpture. *Representations*, 17: 28–61.

Kant, Immanuel 1952 [1790]: *Critique of Judgement*. Translation by J.C. Meredith. Oxford: Clarendon Press.

Kristeva, Julia 1982: *Powers of Horror: An Essay on Abjection*. New York: Columbia University Press.

Kwon, Miwon 2002: *One Place after Another: Site-Specific Art and Locational Identity*. Cambridge and London: The MIT Press.

Lefebvre, Henri 1991 [1974]: *The Production of Space*. Malden: Blackwell Publishing.

Liljefors, Max 2012: Neuronal Fantasies: Reading Neuroscience with Schreber. In: Max Liljefors, Susanne Lundin & Andréa Wiszmeg (eds.) *The Atomized Body: The Cultural Life of Stem Cells, Genes and Neurons*. Lund: Nordic Academic Press.

Liljefors, Max, Susanne Lundin & Andréa Wiszmeg (eds.) 2012: *The Atomized Body: The Cultural Life of Stem Cells, Genes and Neurons*. Lund: Nordic Academic Press.

Lindberg Christensen, Lars, Olivier Hainaut & Douglas Pierce-Price 2014: What Determines the Aesthetic Appeal of Astronomical Images? *CAPjournal* (Communicating Astronomy with the Public), 14: 20–27.

Lock, Margaret & Vinh-Kim Nguyen 2010: *An Anthropology of Biomedicine*. Oxford: Wiley-Blackwell.

Micheva, Kristina D. & Stephen J. Smith 2007: Array Tomography: A New Tool for Imaging the Molecular Architecture and Ultrastructure of Neural Circuits. *Neuron*, 55(1): 25–36.

Micheva, Kristina D., Brad Busse, Nicholas C. Weiler, Nancy O'Rourke & Stephen J. Smith 2010: Single-Synapse Analysis of a Diverse Synapse Population: Proteomic Imaging Methods and Markers. *Neuron*, 68(4): 639–653.

Nagel, Thomas 2012: *Mind and Cosmos: Why the Materialist Neo-Darwinian Conception of Nature is Almost Certainly False*. Oxford: Oxford University Press.

Nisbet, James 2014: *Ecologies, Environments, and Energy Systems in Art of the 1960s and 1970s.* Cambridge and London: The MIT Press.

The Poetic Edda 2004: Northvegr Edition. Edda Sæmundar Hinns Froða. The Edda Of Sæmund The Learned. Translated from the Old Icelandic by Benjamin Thorpe. Lapeer: The Northvegr Foundation Press.

Riegl, Alois 1929 [1899]: Die Stimmung als Inhalt der modernen Kunst. *Gesammelte Aufsätze.* Ausburg: Benno Filser.

Rose, Nikolas 2007: *The Politics of Life Itself: Biomedicine, Power, and Subjectivity in the Twenty-First Century.* Princeton: Princeton University Press.

Rose, Nikolas & Joelle M. Abi-Rached 2013: *Neuro: The New Brain Sciences and the Management of the Mind.* Princeton: Princeton University Press.

Sawday, Jonathan 1996: *The Body Emblazoned: Dissection and the Human Body in Renaissance Culture.* London: Routledge.

Spivak, Gayatri 1998: Cultural Talks in the Hot Peace: Revisiting the 'Global Village'. In: Cheah Pheng & Bruce Robbins (eds.) *Cosmopolitics: Thinking and Feeling beyond the Nation.* Cultural Politics Volume 14. Minneapolis and London: University of Minnesota Press.

Stafford, Barbara Maria 1993: *Body Critics: Imaging the Unseen in Enlightenment Art and Medicine.* Cambridge and London: The MIT Press.

Verschaffel, Bart 2012: The World of the Landscape. *CLCWeb: Comparative Literature and Culture,* 14(3): 1–11.

Image sources for figure 3.2 and 3.3

Figure 3.2. Montage of nine bioscientific images, slightly cropped.
Upper left: Fluorescence micrograph of neurons in culture. Credit Ludovic Collin, 2005. Wellcome Image Awards 2006, http://wellcomeimages.org/indexplus/email/260020.html

Upper middle: Scanning electron micrograph of a kidney stone removed from Kevin Mackenzie, the creator of the image. Wellcome Image Awards 2014, http://www.wellcomeimageawards.org/2014/kidney-stone

Upper right: Molecular model of a bacterial ribosome showing the RNA and protein components in the form of ribbon models. Credit MRC Lab of Molecular Biology, Wellcome Images, 2006. Wellcome Image Awards 2008, http://wellcomeimages.org/indexplus/email/260019.html

Middle left: Colour-enhanced photomicrograph of periodontal bacteria. Credit Derren Ready, Eastman Dental Institute. Wellcome Image Awards 2011, http://www.wellcomeimageawards.org/2011/periodontal-bacteria

Middle middle: Computer simulated image of synthetic pyramidal neurons. Credit Michael Häusser and Hermann Cuntz, UCL. Wellcome Image Awards 2011, http://www.wellcomeimageawards.org/2011/pyramidal-neurons

Middle right: Confocal micrograph of a 3T3 fibroblast cell before it divides in culture. Credit Dr David Becker, Wellcome Images. Wellcome Image Awards 1997, http://wellcomeimages.org/indexplus/email/260014.html

Lower left: Scanning electron micrograph of osteoporotic bone. Credit Professor Alan Boyde, Wellcome Images. Wellcome Image Awards 2002, http://wellcomeimages.org/indexplus/email/260018.html

Lower middle: Computer 3D reconstruction of a mouse embryo at the blastocyst stage, based on multiple confocal micrographs. Credit Agnieszka Jedrusik and Magdalena Zernicka-Goetz, Gurdon Institute, Cambridge. Wellcome Image Awards 2011, http://www.wellcomeimageawards.org/2011/blastocyst-embryo

Lower right: Video still from array tomography 3D reconstruction video *Machinery of Mind*. Credit Smithlab, Stanford University, https://www.youtube.com/watch?v=NFaKLNoI1cA

Figure 3.3. Montage of nine astronomic images, slightly cropped. Credit: European Southern Observatory.

Upper left: Antennae Galaxies composite of ALMA and Hubble observations. Credit: ALMA (ESO/NAOJ/NRAO). Visible light image: the NASA/ESA Hubble Space Telescope, https://www.eso.org/public/images/eso1137a/

Upper middle: VISTA's infrared view of the Cat's Paw Nebula. Credit: ESO/J. Emerson/VISTA, https://www.eso.org/public/images/eso1017a/

Upper right: The Helix Nebula imaged with the Max-Planck Society/ESO telescope 2.2. Credit: ESO, https://www.eso.org/public/images/eso0907a/

Middle left: The Tinker Bell Triplet, by ESO Very Large Telescope/Hubble Space Telescope. Credit: ESO, https://www.eso.org/public/images/eso0755a/

Middle middle: Thor's Helmet Nebula imaged with Very Large Telescope FORS2. Credit: ESO/B. Bailleul https://www.eso.org/public/images/eso1238a/

Middle right: The Dumbbell Nebula, by Very Large Telescope FORS1. Credit: ESO, https://www.eso.org/public/images/eso9846a/

Lower left: A 340-million pixel starscape from Paranal. Credit: ESO/S. Guisard, https://www.eso.org/public/images/eso0934a/

Lower middle: Very Large Telescope image of the cometary globule CG4. Credit: ESO, https://www.eso.org/public/images/eso1503a/

Lower right: Very Large Telescope image of the Arches Cluster. Credit: ESO/P. Espinoza, https://www.eso.org/public/images/eso0921a/

4. Diffractions of the foetal cell suspension: Scientific knowledge and value in laboratory work

ANDRÉA WISZMEG

In the quest for a cure for Parkinson's disease, scientists have travelled many avenues. One is the use of cells from aborted foetuses. These cells have been proven to restore the lacking dopamine production in the brain of the afflicted person. In order to place the cells inside the patient's brain, a so-called cell suspension must be made and administered, which is a liquid produced in a laboratory containing mainly foetal brain cells. This can be transplanted either into rats for research or into human subjects for clinical trials, and theoretically for treatment.

For people encountering the cell suspension, it enters their lives in different ways, giving it diverse shape and meaning; it also gives rise to many different kinds of expectations. Thus, a delicate issue such as the use of the cell suspension, with its foetal origin, provides a good basis for discussing what I would call 'science's understanding of/engagement with knowledge'. Normally when issues of participation in science are discussed, it is done in relation to how stakeholders and otherwise affected people such as for example patients and relatives understand science. The so-called 'information deficit model', where the public was seen as lacking in knowledge and understanding, was a concept common in the research field of the 'public understanding of science' (Evans & Durant 1995; Sturgis & Allum 2004). This model was gradually replaced by views in which engagement and information exchange were regarded as more of a two-way communication between researchers and the public; lay-people's understanding was also seen as a kind of knowledge. The development of a more reciprocal view of knowledge is expressed in the newer concept of 'public engagement with science and technology' (see e.g. Stilgoe, Lock & Wilsdon 2014). Still, even with the newer terminology and the ideas connected with it, much of the focus is on the 'recipients' of scientific results, and less on those who produce them. This is problematic, because it sets the researchers' views apart as something largely free from values, meaning and desire, as opposed to the afflicted lay-peoples' views. In this chapter, I will give a more nuanced and complex picture of how scientists value what they do. With the help of interviews with two laboratory researchers, I focus upon how they understand, value and provide meaning to the foetal cell suspension that they work with. I argue that they do it differently, depending on how they interact with the suspension. The aim is to gain a better understanding of how the scientific knowledge comes into being in a scientific laboratory. The analysis thus problematizes how participation can be understood in a laboratory context.

A diffractive approach

This chapter is based on two semi-structured interviews with two junior biomedical researchers, which I call Emma and James (the researchers have been given aliases in order to protect their anonymity). The interviews focused on the knowledge- and object-production in their respective tasks in a large-scale clinical trial and were based on my observations of their work conducted prior to the interviews. The biomedical project that they are involved in aims to make a definite and concluding transplantation trial with foetal cells from aborted embryos. Both researchers work with the stages of the trial prior to the clinical study. Emma works in the cell laboratory dissecting embryos and preparing the cell material, and James works with transplanting cells to rats and evaluating the cell growth and innervation in them, but he also has experience of working in the cell laboratory. I could thus pose questions to both about the origins of the foetal material and its transformation in the laboratory. I wanted to keep the conversation close to my participants' own practical experience, in order to gain a better insight into the way this shapes their understanding and how they make sense of the foetal cell suspension, both in abstract and material terms.

The interviews were conducted under somewhat different circumstances. I conducted the first interview with James in English, in a quite empty, but because of background music, noisy, café. This interview was the more extensive of the two and lasted slightly more than two hours.

As will be shown and discussed, it was also more elaborative concerning issues of knowledge creation. It seemed as if James and I were more aligned with each other in researching the philosophical issues of science that I was addressing – or at least most oriented towards understanding each other's understandings of the issues at hand. The interview with Emma was conducted in Swedish in a meeting room in my office building, also over coffee; it lasted about one hour and a half. I have translated the excerpts from the interview with Emma used below into English. The non-neutrality of the environment may have impacted the interview situation. However, since I had been the stranger in their laboratories and other premises previously, I believe that the shift in environment into my comfort zone of research ideals and practices may have had a balancing impact on our discussions.

The analysis that follows is an experiment and an attempt at making a shift from a reflexive perspective to a diffractive approach in understanding the interviews. Diffractive analysis is a philosophical approach that focuses on practices that take differences into account and considers how differences are made, while reflexivity is generally more centred on sameness, in seeing the self in the other, and vice versa. The difference between the two research approaches might, however, be minimal. Dialogues, interviews, observations and other fieldwork practices are situated in the world and are regulated by the same natural laws and social codes, no matter what philosophical approach you choose to work with. Still, I argue, that a shift in perspective from reflection and likeness to diffraction and difference, may enable different kinds of knowledge. As educational scholars Alecia Jackson and Lisa A. Mazzei have pointed out concerning how to work with empirical material diffractively, it is not what makes something different, but rather *what* difference is being made, that is of importance (2012: 122). A researcher's subjectivity, for example, is treated as a set of linkages and connections with other things and bodies (p. 135). Rather than it being inherent in a subject, it is highly situational and fluid, with varying durability.

My starting point is an example used by feminist (quantum) physicist and philosopher Karen Barad to explain different kinds of 'research modes'. She borrowed the example from the Danish physicist Niels Bohr. If a person in a dark room holds a cane, the person can intra-act with the cane in two mutually exclusive ways. By holding the cane firmly, the person can use it to navigate the room; the cane then essentially becomes an extension of the subject. If the person instead holds the cane loosely, its features can be examined, turning the cane into an object of study (Barad 2007: 154). The metaphor can be interpreted as indicating two ways of research: either subjugating those studied to the researcher's critical gaze, or enacting the social aspects of the setting collaboratively and letting ourselves as researchers be critically examined by our participants. This takes into consideration how the participants hold, in a metaphorical way, the ethnographer firmly or loosely, but also what kind of knowledge they gain by doing so and what they can set in motion. If we presuppose a boundary between the ethnographer and the 'other', we should remember that the ethnographer is not only holding, but is also being held. Much like the ethnographer, the 'other' will use the research situation to explore the world surrounding them, together as well as separately. The researcher, too, will be the researched.

The question of what happens when encountering and interacting with people in fieldwork, and what our responsibility as researchers should be, is an old theme in ethnology, sociology and anthropology, most recently linked to the question of reflexivity (e.g. Clifford 1986; Ehn & Klein 2007 [1994]; Davies 2008; Gunnemark 2011). Reflexivity has been regarded as a necessity when reflecting upon closeness, distance, likeness, difference and the cementing, or challenging, of power hierarchies and structures between researcher and participant. This was, not least, a welcome change after the colonialist mindset in much of the previous anthropological research.

However, there are limitations. Reflexivity as a philosophical concept is based on the optical metaphor of reflection. It thus sustains a strong subject/object divide, and brings to mind the reflection of oneself in a mirror. And even if one believes that social and cultural research *is* close to 'the world' and performed in proximity with objects and subjects, a reflexive outlook in a sense deems this situation in itself problematic, and calls for distantiating analytical practices and for rational, cognitive reflection. I consider a diffractive approach to be a fruitful extension and development of the reflexive project, by changing our analytical point of departure from 'how can we, being inherently different, understand each other?' to 'what makes us different, when we are originally the same but ever changing and differentiating?' (Mellander & Wiszmeg 2016).

The concept of diffraction is well-known within physical optics (as is reflexivity, arguably). It was introduced to the social and cultural sciences by the feminist scholar Donna Haraway in the 1980s, in connection with her elaborations on situated knowledges (1988). She presented diffraction as an alternative to the – in her view – dissatisfying philosophical concept of reflexivity. Karen Barad then developed Haraway's reasoning on reflexivity. 'Reflexivity, like reflection', Barad says, 'still holds the world at a distance' (Barad 2007: 87). It does not interfere with, but rather reflects the observer like the surface of calm water. Further, it presupposes a pre-existing split between subjects and objects; an a priori division of the world. Quantum physicist Barad could not accept such a division, since to her, existence is always and ever entangled.

While reflection denotes light that is thrown back from objects and returns to its original source in a weaker state, diffraction describes how waveforms spread out and are distorted when encountering objects – thus instead creating *new* beams of light. Waves of water passing through a hole in a dock, or a beam of light passing through a thin slit, will spread out and create patterns emanating outwards. Overlapping, these patterns will in turn create interference, which can be viewed as diversity created in the world through interaction (or, as Barad would have it: intra-action (2007: 139f.)). This might be in physical shape, such as products or artifacts, or as different understandings or opinions of a topic. A change in one parameter of the experimental apparatus, like the distance between the slits, or in the pace you throw rocks into the water, will lead to new patterns and thus different differences.

Diffraction offers other ways of understanding creations of 'I' and 'other'. It differs from the concept of reflexivity, in that it does not deal with reflection or with the mirroring practice of – as Haraway would put it – 'displacing the same elsewhere' (1992: 4). Instead, it deals with interference and the ongoing creation of differences that matter. It does not leave the 'other' as a mere surface of reflection, but lets us think of them as sources of light and makers of waves in and of themselves (ibid.).[1] To create in the world is to make a difference to it; no matter if it is a

---

1. Some attempts to use a diffractive approach on qualitative material have been made in recent years, for instance by Taguchi (2012) and Jackson and Mazzei (2012). The latter use diffraction as a way to map out how the interviewed black woman, who teaches at a white university, may 'intra-act with the materiality of [her] world in a way that produces different becoming' (Jackson & Mazzei 2012: 119). They argue that it is not her being black that diffracts her as different, but that the 'intra-action of bodies, discourses and institutions does so' (p. 125). Blackness, and the concept of race in itself, are enacted by these specific intra-actions.

case of the difference between a beetle and spider, making a political or scientific statement or making a vase in pottery class. All these diverse activities become a question of cutting through the a priori entanglement of the world with the use of an apparatus of knowledge production (Barad 2007: 140). As social and material dimensions of phenomena are understood as entangled, I will subsequently neither make hierarchical order, nor even distinction, between these dimensions of diffractions in my analysis. Interpreting and understanding in discourse with others are actions that create symbolic, social and abstract as well as material phenomena. Divisions taking place and differences being made are *necessary* steps for the forth bringing of the world itself.

Knowledge will thus no longer be understood as the result of reflection, or as stemming from straight lines of sight, but as something emerging through disruptive processes (Mellander & Wiszmeg 2016: 103). It is part of the ethnographers' quest to trace the differences that matter in the subsequent interference patterns. I will in the following present a way of reading the interviews with James and Emma diffractively. Between us, we rendered a number of themes concerning the interpretations, values, knowledge and possible roles of the foetal cell suspension, visible.

Values of the cell

A central aspect of the discussions with James and Emma was the foetal cell suspension as a multi-dimensional object with different kinds of empirical as well as experienced values. The foetal cell suspension is a liquid containing cells from aborted foetuses. What is used is a cell from the developing brain of 6–9 week old foetuses, with the sought-after property of being able to restore dopamine production in the adult, Parkinson-afflicted brain. Some other components are added to the cells in order to keep the cell tissue alive and 'fresh', while preventing further development and division of the cells.

The immediate value of the cells for the researchers that work with them in the laboratory is measured in cell survival (viability) and nerve growth (innervation) of the cells in the brain of the transplanted rats. For the researchers, it is a matter of making this value mobile (Rose 2007), from foetus to patient, via the laboratory (Wiszmeg 2016). This value may even be said to relate to a certain economic value, since it has effects on the researchers' careers and an internal biomedical 'market', even though formal commercial trade with these cells is prohibited, due to them being of human origin, rather than an invention.[2] Aspects – such as the age of the embryos, possible infectious contamination and the amount of surviving cells that each embryo can produce – determine the ultimate value of the foetal tissue. Another circumstance is that the demand for foetal material needed to carry out research and transplantations is greater than the supply; there are never enough aborted embryos available.

With the help of some value categories from moral and value philosophy and ethics, I will now address different values perceived of and attributed to the foetal cell suspension by those interviewed. Highly simplified, something that is *morally good* is usually attributed to humans or human actions, while a *natural good* is something that is seen as a property of objects.

Some different ways of making sense of the foetal cell suspension can be traced in the interview with James. His main task is the transplantation of the suspension to rats in order to evaluate the effect of the cells. A multi-layered discussion on the contents of the cell suspension evolves, spanning from its physical to its existential and symbolic content. This points to the kind of diversity in understanding and rendering of the cell suspension that I want to capture and discuss the implications of. It is clear that James does not have one, but many views of the foetal cell suspension. My first question concerns what James considers the cell suspension to consist of.

JAMES: I guess if I take it from a very physical point of view, it contains disassociated cells that we've collected from foetal tissue. And, you know, going up the scale in that way then it's essentially like a processed piece from abortion material. And, in one way if you think of it – if you were a patient enrolled in the procedure – then you'd see it as a kind of a hope. It's a hope to, you know, something that will modify your life, at least. If that's positive or negative, but you have the hope that it's positive, so, I guess it's kind of transformative. Depending on what point of view.

ANDRÉA: Of course. It has many dimensions, I guess?

JAMES: Yes! But I think for me when I see it or consider it, I just think of the cells and where they came from.

---

2. See Directive 98/44/EC of the European Parliament and of the Council of 6 July 1998 on the legal protection of biotechnological inventions, here: http://eur-lex.europa.eu/legal-content/EN/ALL/?uri=CELEX%3A31998L0044 (accessed 29 July 2016).

When James and I discuss the contents of the foetal cell suspension, he acknowledges, in a sense, the values ascribed to it as both a moral and a natural good. We know that the tissue has therapeutic potential. This also adds an instrumental value (Schroeder 2016); James argues that its instrumental good as being therapeutically promising can be seen as a hope for a cure for the patient. This highlights a transition from a natural to a moral good.

This is, however, a two-faced potentiality. Not only has the cell suspension the possible capacity of curing Parkinson's disease in a person; its original foetal cells would have – if left to develop in utero – made a new human life possible. The two different potentialities could be seen as two kinds of inherent natural good of the foetal cell. If an embryo is aborted it cannot become a baby, but can theoretically be transplanted into a Parkinson patient. Either choice is mutually exclusive and can become the topic of an ethical debate. This, I argue, simultaneously makes the natural good of the cells into moral good, as well.

However, as the foetal cell per se (outside an embryo) would not exist as a proper object without human intervention, it is debatable whether the cell could possibly have natural value disconnected from a moral value at all. Since the existence of the cell suspension depends on human intervention, it seems to escape exact definition on the scale between human action and material object. Not only its constant state of transition, but also its high symbolic value as being transformative, adds to this elusive condition. I would argue that the idea of an intrinsic or inherent good of the foetal cells is a fallacy, since it is based on a projection of our human hopes, dreams and desires. So philosophically and theoretically, the inherent natural good is (in this case, at least) nothing more than an instrumental moral good in disguise.

Still, the idea of an inherent natural good in the foetal cells functions well for pointing out how different values of the cell suspension are perceived by the humans – in this case the biomedical researchers – who are interacting with it. The distinction between inherent natural good and instrumental moral good (no matter how impossible) also helps to visualize the transformative force of the cell suspension – material as well as symbolic. Not only does the perceived value of an object differ; the different knowledge and attributed value renders the object in *itself* different. Taking Parkinson's disease as an example; the knowledge about the disease is not the same for a patient as it is for a relative, researcher or for a medical doctor, neither is the foetal material nor the subsequent cell suspension in themselves. This is a point that James also acknowledged.

The interview with Emma was different. We both seemed aware of the limits of each other's knowledge and philosophical viewpoints, but somehow we lacked the involvement or tools to enable new diffractions together, that is, to explicitly acknowledge the complex plurality of the foetal cell suspension. We rather, but in quite a dispassionate way, registered each other's expressions and opinions. The interview mainly traced some already existing diffractions of the cell suspension, and made them visible. Our difficulty in making progress in the conversation is palpable in the quote below concerning the material components and possible philosophical dimensions of the foetal cell suspension. I entered this theme by asking Emma what she considered the foetal cell suspension to consist of:

ANDRÉA: So then I'd just like to ask very openly, from your perspective, what this cell suspension consists of?

EMMA: Well, I …

ANDRÉA: It's a tricky one! And you can answer whatever you feel like.

EMMA: I just see it as being living cells. – In short.

ANDRÉA: Yes, but when it comes to, I'm thinking purely materially, the substance surrounding it, the hibernation fluid and stuff like that. Is that something you think about, or do you think of it as purely cells?

EMMA: Yes I do. I really don't consider the other.

From here on, we did manage to advance a little further in our common analysis of the pluralities of the foetal cell suspension. However, we never really seemed to enter a sphere of conversation where the level of common analytical understanding or interest could be taken for granted or was a given. Later in the interview, Emma had clearly attained a picture of where my analytic interest resided and could therefore more readily answer my questions. However, this did not mean that she took an obvious interest in these issues herself, as we shall see further on.

The potential in the location of the foetal cell

My idea was to understand if and how objects and experiences are produced plurally in the laboratory setting, thus problematizing the question of which knowledge that is valuable, to whom and in which ways.

In order to get a dialogue going on this plurality or multiplicity, I started a conversation with Emma on the potential of the foetal cells making contact in the new host brain and starting to produce dopamine. This is the mechanism that could help alleviate symptoms of Parkinson's disease. In this way, I tried to pursue my research interest in whether this potentiality could be said to be inherent in the cells, or is created in the laboratory: I think that the potentiality is there, of course, but if the cells were to develop without making contact with a brain, then something else would have happened to them.

> EMMA: Yes, it would, exactly. Just as we have been using this kind of tissue to place in petri dishes. Then they develop into nerve cells at that point, and they don't have the ability, or well, they have lost that potential. Because then they have developed to nerve cells outside a patient's brain. [...]
>
> ANDRÉA: But they have still fulfilled that potential, not in a patient or in a rat, but they still made these nerve cells.
>
> EMMA: Yes, but they lost this entire ... no contact has been made between them.
>
> ANDRÉA: No, of course not.
>
> EMMA: So in that way they might not get the exact same properties.

The potentiality of the cells seems to change depending on context. While Emma and I expressed ourselves differently, it seems that we agreed that the potential of the cells was wasted when left to develop in petri dishes – even if I maintained that they had fulfilled their potential before it was lost. I argued that their potential or natural inherent good has been expressed. However, it is (as addressed above) debatable whether this potential can at all be regarded as natural, given the need for human intervention for it to be transferred from an aborted embryo to the brain of a Parkinson patient. James, on the other hand asserted that the potential must be fulfilled in a certain way and in a certain location in order to be valid. Thus, the only value of the cells is that of their instrumental moral good. The cells are solely a means to an end. In a sense, this is the basis of action for all biomedical research using living tissue.

The cells, the babies and the patients

Emma tells me that she would later like to work in an assisted reproductive technique clinic (ART), in order to establish a kind of 'balance' for all the aborted embryos she has handled. Ethnologist Susanne Lundin has reported about a female scientist working with creating stem cell lines from aborted foetuses, who did not find her work problematic until she became pregnant herself (2012: 22). Emma, on the other hand, did not find the work with foetal cells in itself emotionally disturbing during her own pregnancies. Nevertheless, a sense of guilt seems to have been produced in her, resulting in a desire to make amends at a later date. However, we need to be careful when drawing conclusions about the inner motivations of our participants. The only thing that I can be sure of, is that Emma does express this desire in the interview, and that it makes her diffract the foetal cell suspension somewhat differently than James does:

> ANDRÉA: I just feel it would be strange not to address it, because I think it must affect you both [Emma as well as James], in your professional roles in some way, how you think of it, how you talk about it?
>
> EMMA: Yes, and I kind of feel that after this project, that I would love to work in IVF [in vitro fertilization].
>
> ANDRÉA: Yes?
>
> EMMA: And in some way, I feel that I want to, maybe, be on 'the other side' for a bit! To actually help those who want kids (laughs), instead of those who want to remove them all the time. You just feel that it's so many, I don't know, I have been working with this now for, is it five, six years, with embryos. I mean how many aborted foetuses haven't I handled?

It can be argued that it is the instrumental framing and 'industrial' use of the embryos that makes Emma experience the demands and expectations of her role as a woman and a mother in stark contrast to the demands and logics of her work. She seems to experience the different values (natural or moral) of the foetal cell suspension to be in conflict. Someone without Emma's experiences of being a woman, mother and biomedical researcher, someone outside these intersecting categories, may well have had the same opinion, even if they are not based on the same experiences. Most important for the purpose of this chapter, however, is that Emma's reasoning opens up for thinking about aborted embryos as potential children. Not these specific embryos – since the decision of these women to have an abortion was in no way controlled by Emma or by the clinical trials – but embryos in general.

The potential of the foetal cells to develop into an embryo and subsequently a baby, provided that they stay inside a womb of a woman, is another kind of inherent good of the cells. It can be regarded as a moral good, as well as a natural good. Since the foetal cells do not need human intervention in order to develop into an embryo in utero, this potential seems more obviously a natural good, than the potential of the cells to treat or cure Parkinson's disease.

This way of diffracting the embryo into an object with different biological and therapeutic potentials makes it into, what can be described as a 'bio-object' with transgressive qualities that challenge notions of life, death, of the biological and the mechanical, and of what is ethically defendable, possible and what is not (Holmberg & Ideland 2012).

The social sex system of humanness

These multiple meanings of what the embryo is, comes to the fore now and then in the interviews. James tells me that the instances he recalled as the most unpleasant ones in embryo dissection occurred when the embryos had developed a bit too far and you could define – and here he stresses – their 'physical sex'. It seems like he wants to make sure that I understand that, to him, social sex or gender does not necessarily correlate with biological sex. Neither does he view the biological, or physical sex, as essential. James explained: 'When it's [the embryo] you know, older age and you can determine like physical sex and things like that.' Interestingly, Emma, too, mentions the ability to distinguish the sex of the embryo as an instance when dissecting becomes uncomfortable. She elaborates a bit more:

> But when we did experimental [research] earlier, we accepted more of 'well we don't really know what the ultra sound said, but we just got something here now', and then we said 'ok, but let's take it and see if we can use it'. And then in comes something that is way, *way* too old, and you can almost see the sex and … Well you know, then it's very unpleasant I think, since I'm not used to that. And I mean I sit there with my small instruments, these are small embryos, that you are supposed to dissect – suddenly you get something this big that I can't even use my tools, instruments, on.

Sociologist Bruno Latour has argued that it is only when things break or stop working, or when results are inconsistent, that it is necessary to open the opaque 'black box' of science (1999: 184f.). It is not until then that the mechanics, or the practice that we habitually engage in, becomes visible or problematic for us. The instance when the embryo all of a sudden is too large to dissect with the usual instruments is an example of this unveiling. The situation works as a lens making the foetal material diffract in otherwise invisible ways.

For Emma and James to be able to determine the sex of the embryo means that it has developed further than it should have for their purposes and must thus be discarded for use in the clinical trial. But what they say also implies that (physical) sex is a highly important marker for regarding an embryo as a human being. In my view, it is also illustrative of the extent to which anthropocentrism – the norms and hierarchy that set humans apart from other living beings – rests upon the binary sex/gender system. Although we humans extend this duality to the rest of the animal kingdom, I am not sure my participants would have reacted in the same way if they were dissecting pig embryos. I cannot imagine them feeling a sudden unease on the basis of discovering that the pig embryo could have been a sow or a boar. But then again, they might well have.

When the physical sex of the embryo is distinguishable, it affects how the foetus is diffracted and would subsequently also affect the diffraction of the foetal cell suspension. James' remark concerning this highlights the humanness of the embryo even more, since it implies that it may also have a gender – a *social* sex, not necessarily correlating with the physical sex. We (humans) do not normally attribute gender to (other) animals. The remark is an extension of the insight that the cells come from embryos that might have been babies, which would have been human beings. Thus, James and Emma's unease at discovering the over-development of the embryos can be seen in the light of the inherent (moral) good of the cells to create human life.

Professional roles, conceptual mix-ups

In addition to the entangled differences – or intersections – between James, Emma and myself concerning the diffraction of the foetal cell suspension, there seem to be professional and role based differences within the biomedical project that they are involved in. Consequently the idea of the foetus and the foetal cell suspension is diffracted and becomes plural. This is of course highly relevant for how knowledge and objects are produced. I asked James and Emma where they see the largest discrepancies in the project concerning what to call the foetal material, and if its name changes even if it keeps its material properties. They both stated a number of instances of such discrepancies and began to elaborate on why it may be so. Their explanations related to differences in professional practice, where the greatest difference occurred between staff who were in contact with the foetus as an entity, and those who were not. In addition, staff at the abortion clinics have a relation to the aborting women, which influences the language used and the material reference points that are made.

ANDRÉA: Do you follow me; that it kind of changes names, depending on who's talking, and it doesn't have to be just the tissue, it can be something else.

EMMA: Well, yeah it might actually change, as it goes. I feel that, sometimes I might call it more what it really is. I might say 'the aborting women' and 'embryos', more than maybe the others do at the meetings.

ANDRÉA: What roles do the others that you think of, have?

EMMA: I think of another colleague for example, he might call it 'tissue' and 'donors'. I mean you keep it a bit more … The people at the abortion clinic, they don't say … they are 'patients', they see the patients more clearly.

ANDRÉA: What do they themselves call the tissue when you come to pick it up?

EMMA: They don't say 'tissue'. They would say 'embryos', there.

ANDRÉA: So it's a bit more …

EMMA: … what it is at the time, kind of.

James, on the other hand, found that this conceptual mix-up did not always depend on contact or not with the foetus. He said that he too sometimes would make such mis-references. Thus, the different terms that initially referred to different states of processing of the foetal material are often used interchangeably. However, it may make the staff draw faulty conclusions about the availability of the foetal tissue, since not all aborted tissue is actually usable.

JAMES: Usually I'd say 'the tissue piece', but it's an embryo. You know? I think that's a pretty clear discrepancy. I might call it one thing, but in actuality I'm going to be using specific incorrect language. It'll be something else. You'd say tissue or something, rather than like: embryo. Some people might say 'products of conception'. Or 'abortion material' or whatever, you know. We usually say like 'tissue' or 'embryo', and it doesn't trigger as much sensitivity, if you can say that. So that's kind of an example.

ANDRÉA: That's a clear example as I see it.

JAMES: So I guess I'll usually refer … I mean like today we got these two patients who consented, in that case it is *patients who consent*, you know, and some people will go 'okay, well there's two patients; we'll have two embryos'.

There is thus a re-occurring mix-up between concepts such as donation of aborted foetus, collection of it, an embryo, the dissected tissue, and their material referents, which seems to be influenced both by personal and professional practice, as well as national and cultural attitudes towards abortion.[3] Phrasing and interchangeable use of different concepts is, in short, highly dependent on experience and expectations. Such discrepancies in how the cell suspension is rendered occur between researchers and patients but also as we can see here, among different researchers. Not only is the value and meaning of the cell suspension fleeting; so are the conceptual boundaries of the material needed to create the suspension. There seem to be no exact and solid borders between the aborted foetus, the dissected tissue piece containing the brain cells, the isolated cells themselves and the final cell suspension.

James also told me about how a colleague of his mixes up what it means to have received, on the one hand, a woman's consent to donate the embryo she is about to abort, and on the other hand, the actual embryo to work with in the laboratory. Consent does not automatically mean that the foetal material will be usable for the scientists. It may be too old and too far developed or too damaged for their purposes. Moreover, the woman may still change her mind and withdraw her consent. Still, this colleague occasionally uses 'consent' synonymously with 'embryo'. The colleague has no prior experience in collecting and dissecting embryos, or making the cell suspension, but is specialized in transplanting the cell suspension to rats. This may, according to James, be part of the explanation of why the colleague continues to mix up the concepts and what they entail expectation-wise. This conceptual mix-up also tells us something about value. It seems that the expected usefulness of the foetal material to a specific profession, makes people favour one concept over another. For those who, as this colleague, transplant cells to rats or patients, the value lies in the possibility to have a suspension to transplant. There is no point for them in making a distinction between consent and embryo, since – as far as they are concerned – an embryo is

---

3. Differences in phrasing, depending on for example national research setting, were addressed by James in our conversations, but will not be referred to in ad verbum or quoted here in order to protect the anonymity of others.

equal to a realized consent, when it reaches their hands. Consent is, to them, an instrumental good on the path towards an aborted embryo and a manufactured cell suspension.

Searching together or apart?

I will now leave the analysis of the plurality of the foetal cell suspension and turn to a discussion of if and how the diffractive approach worked to enable the creation of new knowledge in the interviews. We will look closer at the ways in which the approach contributed to a broader and more plural understanding of the foetal cell suspension. The chapter will then conclude with a discussion on how a diffractive approach may be valuable for interdisciplinary research in general, and in particular when biomedicine and neurology are involved.

So, when did we search together and when did we instead search apart? And what consequences did the different modes have for the knowledge brought forth in the interviews? Emma and I projected our respective and different views of the foetal cell suspension onto a common lens – the situation of the interview itself – in order to reach an understanding of the other's views. James and I, instead, first found a common understanding of each other's views of the foetal cell, and then used that common picture as a lens from which a spectrum of other possible cell suspensions were diffracted. We thus spent a lot of time looking for our *common* limits and ways to transgress them. We seemed to align more easily than Emma and I did, and it was easier to exchange and enable new views together around the common topic. Thus, James seemed to have a more flexible way of viewing the foetal cell suspension as plural. He helped me see how the cell suspension was understood and valued differently in different settings of the trial, and how each step was connected, not only to the bringing of the cell into existence, but also to the evidence and legitimacy production around it; that is, what made it 'work'.

I was open about my views and understandings of their knowledge production. James did try to see the issues from my point of view, which was a philosophical and knowledge-problematizing outlook. Even if he did not share my view entirely about the potentiality of the foetal cells, he tried hard to explain the logics of their research so that it made sense to me. He was openly curious about my view and seemed genuinely excited by my way of asking questions.

JAMES: The immunosuppression, I think that's probably the only or the biggest thing, because again, it's a bit of an unknown, you know? You inject it in the belly [of the rats] and then it kind of protects the brain. It's a bit of an unknown process but I think it's kind of crucial. If we had a transplant with no survival, the first question I would get would be: 'well did they get the drugs fully?' So, I think they're kind of essential for preserving or maintaining the integrity of the cells, after transplanting them.

ANDRÉA: But you would still say that, of course you cannot see the drugs present in the brain, but would you still say – '[be]cause they are so crucial for the survival of the cells' – that they are something different than the cells?

JAMES: Transformative?

ANDRÉA: Yes.

JAMES: Yes, it's a different part of the process with the cells, for sure. I think of it as if you make a sand castle, so the cells are the sand castle? And then, the drug is just kind of, to put a barrier to stop the waves from coming, you know?

ANDRÉA: Yes.

JAMES: [Be]cause if you let the waves come, the sand castle falls down and you won't have any neurons, so I guess in that kind of physical representation of it, then they're very different entities. Physically but also functionally.

ANDRÉA: You are painting such nice pictures to me [laughing], I'm getting this visual.

JAMES: I'm kind of making it make sense to myself too.

ANDRÉA: No, but it's good because it helps me understand these things as well. I don't know if you ever get these kinds of questions?

JAMES: No, I don't, so it's kind of fun!

Nevertheless, I sometimes instead chose to talk of the foetal cell suspension in the same way as they are expected to do to in their research environment, that is, either from a strictly cell biological view, or from a broader medical perspective within what can be called an evidence based project. I did this, either by repeating what they had told me about the workings and logics of a practice of theirs, or by trying to draw a conclusion inside their perspectives, based on a piece of information given to me by them in the interview situation or earlier. I did this in order to establish a common base from which to move forward. I have reason to believe that my participants also tried their best to establish such a common base.

In the following section, Emma and I discussed the risks and possible benefits for a Parkinson patient in receiving a cell transplant.

ANDRÉA: I start thinking that it is a bit different then from, I mean that as you say, it is quite a serious operation that is to be done – in very ill patients – but maybe nonetheless with not as severe symptoms. I mean with other kinds of big operations you might – at least as a clinician or doctor – when it comes to other diseases, whatever it might be really, that if someone has serious heart problems you might say that 'ok but it is worth taking this great risk and make a transplantation …'

EMMA: You have a twenty percent chance to be cured, but that twenty, exactly …

ANDRÉA: Is better than …

EMMA: Is better than, maybe, yes.

ANDRÉA: But in this particular case it's special.

EMMA: But in this case it's not, because maybe … Let's say that there's a risk that they develop dyskinesia[4] or something due to the transplant, so that you kind of, might create severe symptoms instead of relief. I mean, that just can't happen, then it's better that this patient is medicated.

I adapted to the logics of Emma's research, by trying to show her that I understood the calculation of risk that they needed to do in order to legitimize the choice of patients for foetal cell transplantation.

In this next extract, James and I talked about the relation between data and evidence. The topic in itself is, of course, very theory-of-science oriented. However, I showed that I tried to enter 'his' perspective and understand the relation in his specific case and the problems and possibilities it may cause.

ANDRÉA: That's really interesting, because this really highlights the relation between data and evidence in a sense. You have the large data set on the patients that you have produced during the trials, and then also, in relation to the end goal, so I think it's really interesting.

JAMES: And we've generated a lot of, I guess evidence, and from my part then with the different tissue. In what ways we collected the tissue, through medical abortions or surgical abortions and then we've also generated a lot of evidence which [does] not necessarily comply with the outcome, but with the variability we discussed in quality, and that's why I said even if it doesn't go to transplanta-

---

4. 'Dyskinesia is a difficulty or distortion in performing voluntary movements' (https://www.michaeljfox.org/understanding-parkinsons/living-with-pd/topic.php?dyskinesia, accessed 25 June 2017).

tion I think that it's not necessarily unsuccessful. Because we've very extensively proven that if you try this in many countries for a long time, this is what you get. And that in itself is a strong body of evidence as well.

The two latter excerpts above are examples of how I tried to search, together with Emma and James, for the premises for knowledge of the foetal cell suspension that their profession requires. It can be argued that this, for a moment, made us more alike. Asking direct questions on knowledge and object production, the way I did in the initial excerpt of this section is, on the other hand, a way of making a difference between them and myself.

When I entered their perspectives, which arguably made us 'more alike', as in the two latter excerpts, this can be regarded as a way of enabling better diffractions later, from a common ground – just as James often took my approach into account during the interview. In a sense, it can be viewed as my way of looking for the right type, strength or grinding of a lens – which would here be their understanding of a phenomenon – in order to later be able to achieve as much diffraction as possible. If it could be established that I understood their point of departure in their reasoning, they would perhaps trust me more as a guide to a philosophical detour, which may then result in a greater conceptual plurality in the ideas concerning foetal cell suspension in our conversation.

James and I seemed to agree – more often than Emma and I did – that our respective different diffractions of the foetal cell suspension could indeed be just one side of the truth. Still, we just as often in a manner of respect kept to our initial standpoints. Whether or not we were sometimes just being courteous with each other in order to keep peace on the issue, I dare not say. However, I would argue that we *did* diffract to a view of the plurality of the foetal cell suspension, by taking the realities we presented to each other seriously and acknowledging them without abandoning our individual points of departure and without aligning fully.

JAMES: Do we modify the foetal cells, or their kind of function or ability? In that way, no. Because I think they can still do the desired characteristics, if I can put it [that way], so they can still, generate dopamine neurons, they can still release dopamine, they can still innervate and go certain areas and survive in the brain. And those kinds of qualities and characteristics we want in a transplantation setting. So again, there's a lot where we don't modify that, but in saying that, if they were left where they were, I mean, there'd be a lot more of them and they'll do it a bit more efficiently, but I don't think we negatively affect it.

ANDRÉA: No. But the function of the cells is …

JAMES: Maintained. I think so.

As to the conversation between Emma and myself – was it more a case of us bonding in a *different* way, rather than us bonding *less* than James and I? Did our way of finding common ground and common understanding of the scientific processes use a different – and on the surface 'non-scientific' – language, which I did not discover at first? Yes, this might actually have been the case. Even if James and I clearly advanced further in our discussion of knowledge connected to the creation and use of the foetal cell suspension, which was what I set out to do, Emma and I instead found a common understanding in discussions of the problems of administration and of the planning of a large multi-site trial.

Conclusions: disagreement in respect

What is the usefulness of a diffractive approach when aiming to understand how scientists work and talk about their practice? It may contribute to developing what I, in the introduction to this chapter, called 'science's understanding of/engagement with knowledge'. The diffractive approach may help visualize the still-not-explicit, as well as the tactile knowledge produced in scientific processes, including those within neurology and biomedicine. A focus on tracking diffractive processes is a way of being responsible for the knowledge brought forth and for the different values attributed to that knowledge. It is a way of accounting for the produced differences and the plurality in objects and artifacts, knowledge, values and experiences, and the different effects they may have on people's lives. As I have tried to show above, a diffractive approach may be especially fruitful within neurological research, since the scientific objects involved – such as the foetal cell suspension or the human brain – are suggestive and multiple. They are at once concrete physical objects (however in varying and changing stages) and abstract surfaces on which dreams, hopes, values and fantasies of life, death and potential are projected. A diffractive approach can help visualize how this happens and by which logic, by allowing and facilitating diversity even at the physical level.

I also think that a diffractive approach could be valuable within interdisciplinary research, since it allows researchers with different epistemological points of departure to assess new knowledge and to identify the potential and force it may have. It may also help to illuminate where and why misunderstandings and misinterpretations between different disciplines occur. As a best-case scenario, it could lead us to acknowledge different expectations and common (but plural) anticipation of research goals and aims in inter- and cross-disciplinary research projects. A willingness to participate, to contribute and to share has an important place in interdisciplinary work and research. It requires openness and a humble attitude towards the other.

References

Barad, K. Michelle 2007: *Meeting the Universe Halfway: Quantum Physics and the Entanglement of Matter and Meaning*. Durham: Duke University Press.

Clifford, James 1986: Partial Truths. In: James Clifford & George E. Marcus (eds.) *Writing Culture: The Poetics and Politics of Ethnography*. Berkeley: University of California Press.

Davies, Charlotte Aull 2008: *Reflexive Ethnography: A Guide to Researching Selves and Others*. Second edition. New York: Routledge.

Ehn, Billy & Barbro Klein 2007 [1994]: *Från erfarenhet till text: Om kulturvetenskaplig reflexivitet*. Stockholm: Carlssons bokförlag.

Evans, Geoffrey & John Durant 1995: The Relationship Between Knowledge and Attitudes in the Public Understanding of Science in Britain. *Public Understanding of Science*, 4(1): 57–74.

Gunnemark, Kerstin (ed.) 2011: *Etnografiska hållplatser: Om metodprocesser och reflexivitet*. Lund: Studentlitteratur.

Haraway, Donna J. 1988: Situated Knowledges. The Science Question in Feminism and the Privilege of Partial Perspective. *Feminist studies*, 14(3): 575–599.

Haraway, Donna J. 1992: *The Promises of Monsters: A Regenerative Politics for Inappropriate/d Others*. New York: Routledge.

Holmberg, Tora & Malin Ideland 2012: Challenging Bio-objectification: Adding Noise to Transgenic Silences. In: Niki Vermeulen, Sakari Tamminen & Andrew Webster (eds.) *Bio-Objects: Life in the 21st Century*. Burlington: Ashgate.

Jackson, Alecia Youngblood & Lisa A. Mazzei 2012: *Thinking with Theory in Qualitative Research: Viewing Data Across Multiple Perspectives*. New York: Routledge.

Latour, Bruno 1999: *Pandora's Hope: Essays on the Reality of Science Studies*. Cambridge: Harvard University Press.

Lundin, Susanne 2012: Moral Accounting: Ethical and Praxis in Biomedical Research. In: Max Liljefors, Susanne Lundin & Andréa Wiszmeg (eds.) *The Atomized Body: The Cultural Life of Stem Cells, Genes and Neurons*. Lund: Nordic Academic Press.

Mellander, Elias & Andréa Wiszmeg 2016: Interfering with Others – Re-configuring Ethnography as a Diffractive Practice. *Kulturstudier*, 1: 93–115.

Rose, Nicholas 2007: Molecular Biopolitics, Somatic Ethics and the Spirit of Biocapital: Inaugural Social Theory and Health Annual Lecture 2006. *Social Theory & Health*, 5(1): 3–29.

Schroeder, Mark 2016: Value Theory. In: Edward N. Zalta (ed.) *The Stanford Encyclopedia of Philosophy*. Fall 2016 edition, https://plato.stanford.edu/archives/fall2016/entries/value-theory/ (accessed 19 May 2017).

Stilgoe, Jack, Simon J. Lock & James Wilsdon 2014: Why Should we Promote Public Engagement with Science? *Public Understanding of Science*, 23(1): 4–15.

Sturgis, Patrick & Nick Allum 2004: Science in Society: Re-evaluating the Deficit Model of Public Attitudes. *Public Understanding of Science*, 13(1): 55–74.

Taguchi, Hillevi 2012: A Diffractive and Deleuzian Approach to Analyzing Interview Data. *Feminist Theory*, 13(3): 265–281.

Wiszmeg, Andréa 2016: Cells in Suspense – Unboxing the Negotiations of a Large-scale Cell Transplantation Trial. *Ethnologia Scandinavica*, 46: 104–123.

5. Mixed emotions in the laboratory: When scientific knowledge confronts everyday knowledge

KRISTOFER HANSSON

This chapter will discuss how researchers, from their base in a scientific laboratory, relate to patients suffering from Parkinson's disease – a disorder of the central nervous system that affects both movement and non-motor symptoms such as mood, behaviour alterations and sleep difficulties. For many patients and their families, contact with scientists involves hope that researchers, within the near future, may offer a new kind of medicine or clinical treatment. There is an expectation that life might develop in another way (cf. Brown & Michael 2003; Hyun 2013). But, as I will discuss in this chapter, this is not something that researchers can realistically promise. I will problematise communication challenges and the mixed emotions that occur in the laboratory when scientific knowledge is confronted with the everyday knowledge and expectations of patients and their families.

Ethnographic methods in the laboratory

I have followed a research group in its everyday working situations and in its various ways of interacting with patients and relatives. The interaction is often handled via e-mails and letters from patients to the research leader. The patients are also present in the laboratory in other and more abstract ways, such as in the form of a small piece from a deceased patient's brain, together with associated test-results in an Excel file. Before I present the different working situations, I will give a short introduction to my ethnographic methods in the scientific laboratory.

Through ethnographic observation, the working conditions and the material objects of the biomedical laboratory may be made visible; this is

95

where knowledge is created, defined as scientific knowledge, in this case concerning the brain and diseases of the brain (cf. Latour & Woolgar 1979; Knorr-Cetina 1999). As sociologist Deborah Lupton argues: 'It is rather to emphasize that material objects, or ideas, are important components of the way in which scientific knowledge and practices [...] come into being and operate' (Lupton 2012: 17). During my months in the laboratory, I studied the core activities of the research. This is where the researchers performed experiments, compiled and analysed data and presented this data in various contexts, such as at conferences, internal meetings and in renowned journals. Such activities created a distinct orientation in the everyday work of the researchers. My aim was to understand and contextualise ordinary events in the day of the researchers, including the actions themselves, the spontaneous conversations and the work with material objects. People often do things, which they may have difficulties in verbalising or reflecting over, for example, in a formal interview (Frykman 1990; Frykman & Gilje 2003). An outside observer might get the impression that nothing much happens in a laboratory. It might be thought that things largely involve waiting for an experiment to be finished, for an article to be sent back from a peer-review or just sitting reading. Cultural practices are created in all this waiting, which therefore is significant to observe and illuminate (Ehn & Löfgren 2010). Nevertheless, events also occur in the laboratory that the ethnographic observer will not so easily understand. During the course of the observation, I therefore found it essential to discuss matters continually and to ask the observed researchers informal questions.

Further material that I have been given access to is letters and e-mail sent to the research group and the research leader from the research communicator and from patients and their families. I was allowed access to this material in the office of the research communicator, where I could read it and take notes. I discuss it in the chapter, but do not quote it explicitly, in order to enable anonymity for the research group as well as for the patients and their families. For the same reason, I have not supplied references to news articles, accessed on the internet, about the results of the research group. I am aware of the fact that the readers will have to trust my analysis without being able to study the background material themselves, but a balance has to be made between a source critical perspective and the risk of exposing this group of people; in this case, the latter consideration was of greater weight.

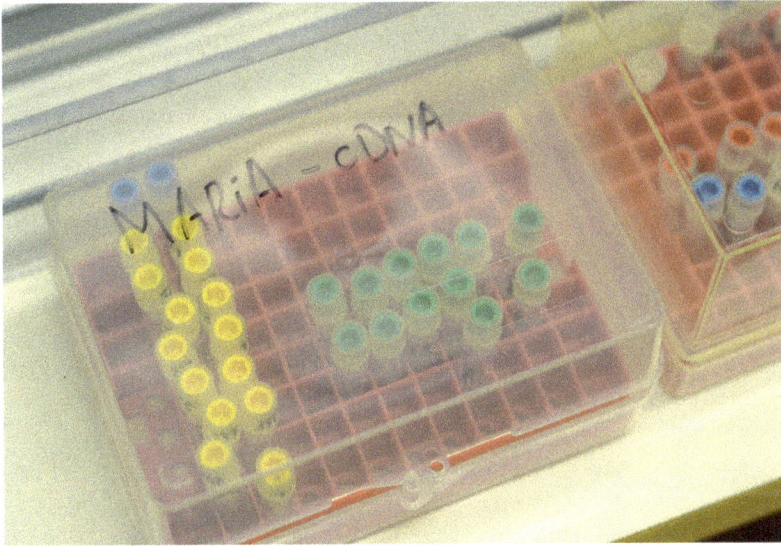

Figure 5.1. Material objects in the scientific laboratory. Photo: Kristofer Hansson.

Scientific knowledge comes into being

One of the days that I visit the scientific laboratory, it is quiet; there are two persons in the two adjoining rooms of the laboratory, as well as myself as the observing ethnographer. This day, I am following Maria, a young researcher working on a postdoctoral project on cellular alterations affecting the brain. In the other room, Anders, the laboratory assistant, concentrates on preparing a number of samples to be part of one of the experiments conducted by the research group. He has the radio on very quietly and the music drifts in to Maria and me in the next room. The most prominent sound in our room is the laboratory fan, but now and again, the built-in motor of the microscope can be heard when Maria adjusts the focus. Sometimes there is the faint click of the computer mouse when she marks the fields for her calculation; she tells me that these represent the number of proteins in a brain sample. Her work is part of a research project with the purpose of investigating the development of Parkinson's disease in the brain's basal ganglia.

In the room, there are several brown cardboard boxes, each box containing white plastic containers. Each box is marked with different numbers denoting the deceased patient. I read: '95-74, 92-49'. These numbers mean nothing to me. Maria tells me that the samples are imported from a brain bank, based abroad, outside Sweden. The numbers on the boxes are related to an Excel-file in Maria's computer with information about the deceased patient. This information enables her to distinguish between samples that she is not interested in and those that she will use. She tells me that a prerequisite for the samples she will analyse is that the patient does not have any other medical issue apart from Parkinson's disease, since this might affect the cells to be analysed.

The brain sample is delivered on a glass slide, which Maria must prepare by applying various techniques, in order to make the dimensions she is interested in visible. The preparations take around twenty-four hours while the sample is treated with different substances. Before Maria then puts the slide into the microscope she draws a line with a black felt-tip marker around the brain sample; a multitude of such irregular figures will be drawn during the day. Maria feeds the slide into the microscope and a picture appears on the computer screen to the right of the microscope. I ask her what we see on the screen and she answers 'blood vessels', pointing to some small dots on the screen.

This account illustrates the day-to-day character of the work which enables Maria to generate scientific knowledge. It is an everyday situation that I understand through the concept of lifeworld, which is based in concretely experienced reality. A lifeworld is both pre-reflexive and pre-scientific, since it, in this case, both precedes the scientific knowledge generated by the biomedical researcher and is a prerequisite for it (Bengtsson 2001). These pre-reflexive and pre-scientific qualities can be described as the intentionality of the researcher – an awareness of something – of the objects, environments and other researchers surrounding the researcher (Husserl 1972 [1913]). This intentionality is, not least, a bodily capacity. For somebody working in a laboratory, it may involve an accomplishment in preparing the samples, adjusting the microscope, making notes of the results produced by the computer programme, writing an article, and so forth. Maria possesses a high degree of what the phenomenologist Maurice Merleau-Ponty would term a bodily accomplishment in handling samples and medical technology (Merleau-Ponty 2002 [1945]). This bodily accomplishment can be said to create Maria's lifeworld in the laboratory.

I will also use the term *bio-objects* to denote what the brain samples become in the laboratory; they are prepared and examined and if everything turns out right they might generate new scientific knowledge about Parkinson's disease and the brain. However, as the sociologist Tora Holmberg and the ethnologist Malin Ideland point out, these bio-objects can also be moved to other contexts where they may challenge traditional, juridical, political, ethical and cultural ordering systems (Holmberg & Ideland 2012). They may become controversial when they move from science into contexts where a greater degree of traditional everyday knowledge prevails, in this case concerning the brain and its cells. A bio-object should thus be regarded as an unstable object, different within the different contexts where it is used and understood, something that I will return to below.

Yet the lifeworld of the scientists is more than just the laboratory, its many objects and researchers. Further, it is the ideas that the researchers have concerning why their research is relevant (Lupton 2012). While I am sitting beside Maria studying her work, she tells me about the purpose of her research. Her hypothesis in this project is the idea that people with Parkinson's disease have more blood vessels than others in their brains. Today she is examining this hypothesis but, as she explains to me, it is linked to a much wider context extending beyond today's monotonous work in front of the microscope. If the hypothesis is verified, she might be able to continue along this line using the specific animal models that the group works with. These involve another kind of bio-object, the transgenic mice that have all developed symptoms similar to Parkinson's disease. She can then apply various medical substances in order to find those that might lead to an alleviation of the disease, or a cure. If positive effects are found in the animal tests, then, Maria says, clinical tests on humans can be commenced. 'But this is a long process, of course', she says. Nevertheless, ideas about the wider relevance of the research are an important dimension of the individual researcher's lifeworld, not least, in the kind of positive feelings associated with their, often monotonous laboratory work.

Much more ambivalent feelings occur in relation to the fact that the samples from the brain bank come from persons with a very serious disease. In Maria's account of her work, there seems to be a strong awareness, not only of the work she is doing today, but also of a much more far-reaching vision, extending outside the laboratory. At the same time as this perspective directs Maria's research and her lifeworld, it also seems to create an ambivalent feeling concerning the fact that it is a long process, which may lead the researcher in the wrong direction. And, even if her aim is to solve the enigma of Parkinson's disease, she is aware that she cannot realistically expect that she will be the person to find the definite cure.

This ambivalence also concerns Maria's own research career and even the work of the group; they have to succeed in finding financial support for their research in a severe competition for limited research funding (Müller 2014). This situation becomes clear in another episode during the day, when laboratory assistant Anders comes along for a chat, and Maria mentions that she has high expectations that the study, which they are working on together, will produce results. They have both spent a lot of time on it and agree that it is slow work. To me, as an outsider with very little insight into the research, an abundant measure of patience, concentration and perseverance seems to be required. The threat that the hypothesis will not be possible to verify is always present as a potentiality. Thus, the feelings present within the researchers' lifeworld go beyond the everyday practical work in front of the microscope; they include expectations and optimism, but also a measure of reserve.

What I have described ethnographically here resembles a formation history in which the researchers enter their laboratory and generate, not only that which *is*, but also that which *comes into being* (Bachelard 1968). As noted above, we can understand this process of coming into being through the concept of bio-object, and as a critique against Immanuel Kant's theory of knowledge. In Kant's theory, we assume that the objects of knowledge must be what they are, they cannot be anything else. They must be identical with themselves in all situations. They should also occur in a specific place, they cannot be in two places at the same time (Kant 1965 [1781]). The concept of bio-object suggests something else, that scientific knowledge, for example concerning the brain, is not a fixed object. Rather, it is created and, in Bachelard's terms, comes into being in a multiplicity of situations, and is different depending on where the bio-objects occur. In the laboratory, the objects are transformed chemically and are analysed; they then move away from the laboratory and into other arenas. The scientific knowledge about the objects is something other than the knowledge that occurs in the lifeworld of people who live with Parkinson's disease, as patients or relatives (Mol 2002). These different forms of knowledge, and bio-objects, are, however, not isolated from each other. They do interact, as will be discussed below, in the public arena, for example in media. Nevertheless, such encounters may also, as we shall see, occur in the laboratory, in the researchers' own lifeworld.

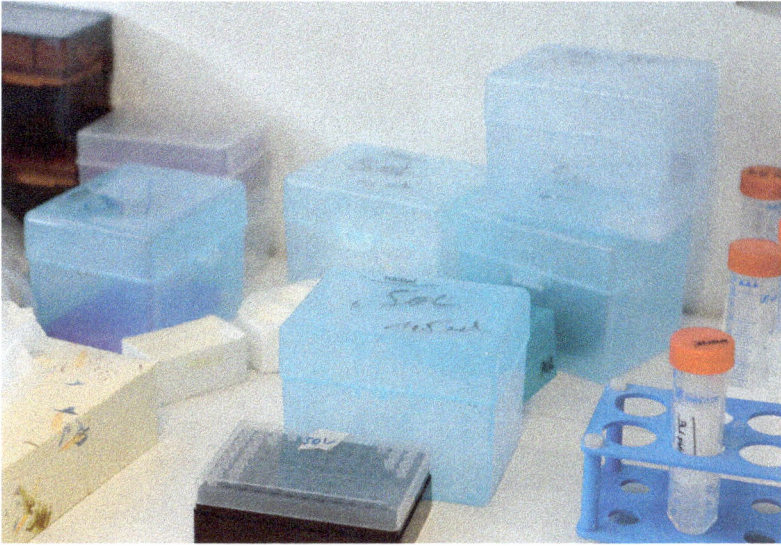

Figure 5.2. It is in this laboratory environment that the researchers are trying to find elements of a cure, hidden among boxes and test tubes. Photo: Kristofer Hansson.

Encountering patients and their families

Biomedical research is often future-oriented in its communication with the public. Not only are the latest discoveries in the laboratory presented – clinical tests or statistical calculations – the results are also associated with prospects of finding a future answer to medical problems or a cure to the disease studied by the specific group. Frequently, the implication in press releases is a feeling of hope for the sufferers – the patients and their families. The discourse gives hope of a possible new kind of medicine or clinical treatment within the near future (Brown & Michael 2003). Not infrequently, this discourse is linked to the necessity for further funding of new studies or clinical tests. The ethnologist Susanne Lundin has metaphorically described this situation as the researcher stepping out, like a fashion model onto the catwalk, and showing the best of what research can offer in the form of future treatment or new knowledge (Lundin 2002). Thus, not only are the results of previous research presented to the external world but they are linked to a vision and hope of the future course of work.

This information may be eagerly seized upon by media and those suffering from the disease, which became evident during the course of my study. The studied research group sent out a press release following Maria's promising results and their publication in a scientific journal. In this press release, the research leader described the discoveries and their implication for the patients suffering from Parkinson's disease. One widely circulated Swedish evening paper then presented a more or less unrevised version of the press release, which was careful in its promises. The medical background of Parkinson's disease was introduced in the article, as well as the number of people who develop the disease and the number of people affected by it. The text also mentions how the experiment was performed. The quotations that occur in the article are from the research leader, and taken from the press release.

In this way, the article follows the logic in which scientific knowledge, as it is presented in scientific articles, is converted and transformed into what I have termed everyday knowledge. There is nothing in the article that the reader cannot relate to, or that can be described as revolutionary information. Instead, the presentation reproduces the knowledge that already exists within the public concerning Parkinson's disease (Hansson 2003). I would argue that a form of epistemological break occurs here where scientific knowledge has been translated into everyday knowledge. Together with the research communicator, who has written the press release, the research leader has found a language different from the scientific language. Consequently, I am critical towards studies that claim that science cannot or will not adapt its scientific language (Wynne 1996; Clark & Murdoch 1997; Pellizzoni 1999 & 2001). In this case, and with the help of a scientific communicator, the message was formulated in a way to be accepted by editors, published as news, and made understandable to a larger public.

Researchers can, however, not control the whole process. In this case, the newspaper themselves chose the headline with an aim to attract readers. In the headline, the word 'hope' was used and it alluded to the prospect of a possible treatment for patients with Parkinson's disease within the near future. In other words, before even starting to read the article, the reader was presented with a message that the research group might soon be able to offer treatment for sufferers. In addition, the paper interviewed a representative of the Swedish Parkinson's Disease Association, who, while emphasizing the need for being careful about using the word 'cure', also mentioned being 'hopeful' that the new discovery would be an effective way of impeding the progress of the disease. This mention may have been based on a statement by the research leader in the press release, that the new results would enable further research leading to new knowledge, which may in turn lead to clinical testing in the near future.

This formulation attracted attention in both national and international media and aroused hope among afflicted patients and their families. It had the effect, during a period of time, that a considerable amount of e-mail and letters were sent to the research group with enquiries about whether it would be possible to participate in the clinical tests. Nobody appears to have understood the phrasing in the sense intended by the research leader: that the results only showed the relevance for further research. Instead, the sentence and the article were interpreted through people's everyday knowledge and were understood as the possible lifeline they had been waiting for.

The head of the research group receives many e-mails and letters from patients and members of their families, enquiring about the progress of the research conducted by the research leader and asking whether it is possible to take part in the clinical trials. Together, the research leader and the research communicator discuss the answers that should be sent to the writers of these e-mails and letters. As the example above of the press release shows, it is a question of careful wording. At one stage in an exchange of e-mails between the research leader and the communicator, the former wrote the following: 'It is hope, they seem to need … although I do not want to make them too hopeful of any kind of miraculous cure.' In this way, she is carefully reflecting upon how she can convey a nuanced message to those with Parkinson's and their families. In the next section, I discuss the difficulties experienced by the research group in responding to such encounters and to the feelings of hope that their results may encourage.

Handling encounters emotionally

With the digital technology of today, researchers are more visible than before. Patients and the public can easily find information about them and their research, and also get in touch with them. The enquiries received via letter or e-mail by the research leader are often very personal. The letters supply an insight into the way the patients and their families formulate their everyday knowledge about Parkinson's disease and its reality for them in various social and personal situations. One way

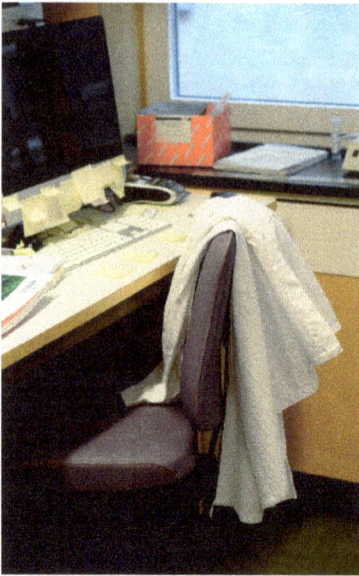

Figure 5.3. A large part of the working day for the researchers is spent in front of the computer. Photo: Kristofer Hansson.

to understand this everyday knowledge is to regard it as based in a body with an existential experience of the changes that it is undergoing (Kleinman 1988). However, it is also a body that is part of a historical situation. This becomes clear, not least in the writers' reactions to the results of the research group; when they take part of these results, they respond to them in different ways.

Some describe how Parkinson's disease has affected everyday life. The e-mails are sometimes short, just a few sentences, but long letters covering several pages are also sent to the research leader. In many letters, the writer describes the course of the disease, like a short case history. The patients often give an account of the medication they have tried throughout the years, the way these medicines have affected the body and their attitude to the medicine. Many people mention the side-effects of the medicine. A specific medicine makes somebody feel ill. Another describes how the shakiness is always getting worse, making everyday situations difficult to handle, like showing a train ticket on the mobile phone or cutting onions when cooking. A changing body may also be difficult to handle in social situations: problems with drooling, balance, voice or a hunched back torment the letter writers in various social contexts. Dementia is a common topic.

In many cases, the letters include a request about being permitted to participate in what they think is an impending clinical study. The research leader tells me that sometimes the afflicted persons themselves make the enquiry and sometimes it is their close family, if the afflicted person can no longer manage to write him- or herself. In all cases, there is a hope that this particular research group will find the solution to a process that gradually is changing the persons' bodies and minds in a way they were not prepared for. Many are profoundly knowledgeable about the medical aspects of the disease, and they know the implications of being a test person in a clinical trial. Most writers express 'having been given new hope' when reading about the new study in the morning paper or seeing it on the television news. They hope to participate in future clinical studies; frequently they write that they would take part even if there are side-effects from the medicine.

A Monday meeting

I now turn to an account of a meeting in the laboratory where one such letter from a patient was discussed, including how to respond to it. The situation provides an insight into the encounter between two kinds of realities: the researchers' experiences of high expectations that contrast with those of the patients, making it challenging for the researchers to respond to the patients' existential and emotional demands.

It is time for a Monday meeting at the laboratory and a group of seven persons are gathered around the table. The research leader is not there this morning, but the meeting is carried out as usual with Maria leading the discussion. First, some of the doctoral students present the work that they will be doing the coming week. Then it is Maria's turn. She tells us that she will be answering some of the e-mails that have been sent recently to the research leader, since the research leader is occupied with a major application for research funds, which needs to be submitted this week. Although it is now a few months ago that their study attracted interest in media, they still receive enquiries from patients and families, and Maria must now answer one such e-mail. She is usually very professional in her role as a researcher, but when she tells the group about this e-mail, she is noticeably affected emotionally. She tells the group that the e-mail is written by a woman whose husband is diagnosed with Parkinson's disease. The woman now hopes that the group will speed up their research because her husband is in great need of treatment. Maria tells the group about her concerns. Even though Maria works in the laboratory and has no daily contact with patients suffering from Parkinson's disease, she has met such patients in many different situations. She knows what the disease does to their bodies and how frustrating it is for the afflicted person, as well as for their near family, when the medication is no longer as effective as before. Now the experience described in the letter appears as a direct entreaty to her as a researcher; she should be the person to help the woman get her husband back to the person he once was, but how?

How is she to describe the research process and the scientific goals to this woman, including the limitations of their results?

Somehow, this situation seems to be too much for this group of doctoral students and they have no good answer for Maria. Instead, they start to joke nervously about what to reply. The jokes are not made at the expense of the woman or the afflicted man. Instead, their restrained laughs arise in relation to the uncertainty about what they, as young researchers, can do for the couple; I perceive this reaction as a sign of helplessness. The woman has read the article about the latest discovery of the research group; understanding it with her everyday knowledge, she sees the relevance of the research – but in the laboratory, this relevance becomes difficult to relate to. For the students around the table in the laboratory, so far from the everyday life of the patients, the woman's hopes become somewhat nonsensical. Their immediate worries concern how their experiments with the transgenic mice should be performed, how they should tackle the critique of their scientific studies, or the circumstance that they are almost finished with their doctoral studies and must start thinking about what to do after their four years of research work. They know that the competition for a position as a researcher is severe and that they presumably will need to leave the university (Müller 2014). In this situation and for this group, any long-term plans to find a cure for Parkinson's disease appear as an almost impossible endeavour. When they, all of a sudden on a Monday morning are confronted with the hopes that their discoveries have roused in the woman who sent the e-mail, they are at a loss.

In other words, the bodily experiences of the patients and their families can be difficult to relate to emotionally for these young researchers. Not because they have emotional difficulties in understanding this experience, but rather that there is another form of knowledge about the disease expressed in these mails, which can be hard to get a grip on from the angle of the researchers. It is not part of their everyday lifeworld in relation to the phenomenon of Parkinson's disease. Somebody even jokingly, and to the tittering of the others, suggests that they should be given more money to speed up the research; I interpret this as a sign of helplessness when confronted with the woman's hopes. But Maria was not amused. Jokes were not the help she needed. Instead she really wanted to understand this other lifeworld and be able to give the letter-writer an honest and nuanced reply this Monday morning.

Conclusion: complex encounters

A point that probably at first does not strike a visitor to a scientific laboratory, is that in these rooms and corridors there is usually a profound sense of the presence of people with a specific diagnosis. My purpose of this chapter has been to illuminate the complexity that occurs in these places, especially in the encounters that the researchers have through e-mails and letters that the patients and their relatives send in. One way of doing this is to describe the laboratory as a place where the scientific knowledge confronts the everyday knowledge that the patients and their families have of a specific illness. From this perspective, it is not only scientific knowledge in the laboratory that comes into being, everyday knowledge is produced too, and operates in different ways. This becomes obvious when the researcher takes the step into the public arena, and encounters are created that require an understanding of the everyday knowledge of the patients and their families. Therefore, to transform scientific knowledge into knowledge that can be understood by the public, the researcher also needs to be able to cope with people's great and small hopes and expectations. In this chapter hope occurred in the many letters and e-mails that were received by the research leader, in which patients and their families hoped that this particular research group would come up with the solution to the enigma of Parkinson's disease. In such a way, hope was constantly at hand as a realistic expectation of the continued development of life for the patients and their families. It is therefore central to problematise the communication challenges and the mixed emotions that occur in the laboratory when scientific knowledge is confronted with the everyday knowledge of patients and their families.

References

Bachelard, Gaston 1968: *The Philosophy of No: A Philosophy of the New Scientific Mind.* New York: Orion Press.

Bengtsson, Jan 2001: *Sammanflätningar: Husserls och Merleau-Pontys fenomenologi.* Göteborg: Daidalos.

Brown, Nik & Mike Michael 2003: A Sociology of Expectations: Retrospecting, Prospects and Prospecting Retrospects. *Technology Analysis and Strategic Management,* 15(1): 3–18.

Clark, Judy & Jonathan Murdoch 1997: Local Knowledge and the Precarious Extension of Scientific Networks: A Reflection on Three Case Studies. *Sociologia Ruralis,* 37(1): 38–60.

Ehn, Billy & Orvar Löfgren 2010: *The Secret World of Doing Nothing*. Berkeley: University of California Press.

Frykman, Jonas 1990: What People do but Seldom say. *Ethnologia Scandinavica*. 1990: 50–62.

Frykman, Jonas & Nils Gilje (eds.) 2003: *Being There: New Perspectives on Phenomenology and the Analysis of Culture*. Lund: Nordic Academic Press.

Hansson, Kristofer 2003: Djur som donatorer: Xenotransplantation i svenska medier 1995–2002. *Lundalinjer*, 120: 1–71.

Holmberg, Tora & Malin Ideland 2012: Challenging Bio-objectification: Adding Noise to Transgenic Silences. In: Niki Vermeulen, Sakari Tamminen & Andrew Webster (eds.) *Bio-Objects: Life in the 21st Century*. Burlington: Ashgate.

Husserl, Edmund 1972 [1913]: *Ideas: General Introduction to Pure Phenomenology*. New York: Collier.

Hyun, Insoo 2013: Therapeutic Hope, Spiritual Distress, and the Problem of Stem Cell Tourism. *Cell Stem Cell*, 12(5): 505–507.

Kant, Immanuel 1965 [1781]: *Critique of Pure Reason*. Unabridged ed. New York: St. Martins Press.

Kleinman, Arthur 1988: *The Illness Narratives: Suffering, Healing, and the Human Condition*. New York: Basic Books.

Knorr-Cetina, Karin 1999: *Epistemic Cultures: How the Sciences Make Knowledge*. Cambridge: Harvard University Press.

Latour, Bruno & Steve Woolgar 1979: *Laboratory Life: The Social Construction of Scientific Facts*. Beverly Hills: Sage.

Lundin, Susanne 2002: The Body is Worth Investing in. In: Susanne Lundin & Lynn Åkesson (eds.) *Gene Technology and Economy*. Lund: Nordic Academic Press.

Lupton, Deborah 2012: *Medicine as Culture: Illness, Disease and the Body*. Third edition. Los Angeles: SAGE.

Merleau-Ponty, Maurice 2002 [1945]: *Phenomenology of Perception*. London: Routledge.

Mol, Annemarie 2002: *The Body Multiple: Ontology in Medical Practice*. Durham: Duke University Press.

Müller, Ruth 2014: Racing for what? Anticipation and Acceleration in the Work and Career Practices of Academic Life Science Postdocs. *Forum: Qualitative Social Research*, 15(3).

Pellizzoni, Luigi 1999: Reflexive Modernisation and Beyond Knowledge and Value in the Politics of Environment and Technology. *Theory, Culture and Society*, 16(4): 99–125.

Pellizzoni, Luigi 2001: The Myth of the Best Argument: Power, Deliberation and Reason. *British Journal of Sociology*, 52(1): 59–86.

Wynne, Brian 1996: Misunderstood Misunderstanding: Social Identities and Public Uptake of Science. In: Alan Irwin & Brian Wynne (eds.) *Misunderstanding Science? The Public Reconstruction of Science and Technology*. Cambridge: Cambridge University Press.

6. Meetings with complexity: Dementia and participation in art educational situations

ÅSA ALFTBERG & JOHANNA ROSENQVIST

In January 2013, a three-year project, 'Meetings with Memories' (in Swedish: Möten med minnen), was launched in Swedish museums with guided tours for dementia-afflicted audiences.[1] The project involved altogether 88 Swedish museums and was headed by the Alzheimer Foundation (Alzheimerfonden). Many Swedish art museums were part of the project, with the purpose that the art educational situations would create participation and dialogue with a group of people usually absent from the museums. Engaging in art is regarded as a possible therapeutic rehabilitation for people with neurodegenerative diseases; dialogue and producing art enhance cognitive abilities and quality of life (Rosenqvist & Suneson 2016).

The therapeutic promise (Rubin 2009) within art pedagogy and the strive for participation and dialogue are part of the context of the art educational situations we have studied. In this chapter, the aim is to explore participation in art educational situations in relation to the target group, people with dementia. How do cultural norms affect the possibilities of participation, and what kind of participation is desirable?

Participation in this context must be examined through cultural beliefs and norms concerning dementia. Creating individual identity and the 'identity work' of modern society is very much recognized

---

1. We use 'dementia' in the text as an everyday inclusive word for a range of different neurodegenerative diseases such as Frontotemporal dementia, Multi-infarct dementia, Lewy body dementia and Alzheimer's disease, mainly focusing on the latter due to the initiative of Alzheimerfonden (http://www.alzheimerfonden.se/motenmedminnen, accessed 19 May 2015).

as a project about cognition and memory, considered as two essential attributes of humanness and equated with selfhood in Western culture (Basting 2003). People with dementia are thus considered to have lost that unique individual personality and identity that define them as humans, since the disease dissolves and disintegrates the selfhood of the individual (Kontos 2004). Dementia is accordingly regarded as a social death, where the affected person is stripped of his or her subjectivity. Participation, on the other hand, is grounded in beliefs of social inclusion and meetings between people who are fully human, active subjects (Seigel 2005). How can we then understand participation in relation to dementia? Can dementia give us a new understanding of participation? Discussing participation and dementia in a societal context of increasing cerebral focus also contributes to illustrate aspects of neuroculture, its impact on perceptions on the human brain and selfhood. Theories about subjectivity and selfhood are an important frame of this chapter and will be developed further on.

The chapter's empirical examples derive from participant observations that were conducted in several different locations. For ethical reasons, we will hereafter refer to these as taking place at the anonymized Art Museum. The setting is this: to participate in 'Meetings with Memories' at the Art Museum you sign up together with a friend, partner or caregiver. You are invited to come on a Monday, which is normally a day when the museum is not open to other visitors. Everyone is given nametags, including the guide or art educator who will show you a selected number of works in the galleries for the next hour or so. These nametags make it possible for us to call each other by our first names. It indicates a difference from regular tours where only guides and artists are named. The focus for our observations at the museum is the participation and dialogue in front of the carefully chosen paintings or sculptures. The dialogue is initiated and managed by the art educator. But the answers to the questions and the free associations and comments are all yours to give. Sometimes you are instructed to discuss with your fellow participants in the group, sometimes you are encouraged to have a dialogue with the art educator. Another part of the context is of course the artworks and the rooms of the Art Museum setting, where the artwork is given some space on either side, to set it aside and avoid visual distractions.

Let us start with an example from our field work of how such an art educational situation may take place:

We are gathered in the entrance hall at the Art Museum, waiting for a guided tour to start. There are only a handful of people who have signed up for this special tour, directed towards people with dementia and their companions: a family member, a care giver or a close friend. The art educator who is in charge of the tour arrives and she welcomes us. She asks for our first names, writes them down on stick-on labels and hands them out for everyone to put their name on their shirt or jacket. Then it is time to take the lift to the floor where the tour is about to take place. A guard joins us, but otherwise the museum is empty since it is closed on Mondays.

When we step out of the lift, the art educator asks us to take a folding chair from a rack placed next to the lift door and we head for the exhibition. We take our seats in front of one of the paintings. The art educator starts to tell us about the work of art, focusing on light and colours. The guard stands unobtrusively behind us. The art educator asks us what we see and what kind of colours we perceive. She then invites us to discuss in pairs the feeling of summer when the sun is shining, like in the painting. Finally, we are asked to come really close to the painting, and then back away and look at it from a distance.

We begin to move to another painting, and at the same time another group of people is passing us. They are going to an arranged lecture in the next room. One of the women starts to follow them instead of the tour group, and her friend goes after her, returning her to our group.

The next painting we stop at is placed in another room, and the acoustics are rather bad. The painting's name is reminiscent of a song and one person in the group starts to sing a few notes. The art educator takes a small loudspeaker from her pocket and connects it to her mobile phone, and plays the song for us. Then we discuss how the impression of the painting is changing, due to the music. Next, the art educator asks us to discuss, in pairs, a favourite place and what colour we associate it with. One woman tells us that she has lived all her life in this town, but her parents came from the countryside and as a child she spent a lot of time there. The art educator asks her about the time in the countryside and the woman declares that white is her favourite colour because it is like snow, and she mentions the games she used to play in the snow as a child.

The Swedish project 'Meeting with memories' has made use of the methods of the 'Meet Me' project at the Museum of Modern Art in New York, featuring guided art tours for Alzheimer patients since 2006 (Rosenberg et al. 2009). An evaluation study of the project from the New York University Excellence Centre on Brain Ageing and Dementia reveals significantly heightened life quality for participants, and very high appreciation by relatives and caregivers (Mittleman & Epstein 2009). This has sparked several follow up projects apart from the one mentioned above (Caulfield 2013; Schall, Tesky & Pantel 2015). For example, Swedish art institution Malmö Konsthall has since 2011 arranged guided tours for people with a neurodegenerative condition, also based on methods from MoMA's 'Meet Me' project, but with the addition of a creative workshop after the tour.[2] Observations by the educators at Malmö Konsthall and caregivers also indicate positive cognitive, social and life quality effects, in line with previously mentioned studies. Creative stimulation is thought to be a positive influence on perceived well-being and everyday function (Holmbom Larsen, Minthon & Londos 2014). These activities are not only designed to benefit the person suffering from dementia. The art therapy experience is also thought to benefit the well-being of caregivers as well, besides increasing the dementia patient's awareness of social support and feelings of connectedness with others (Reid & Hartzell 2013).

Historically and today, educational programmes in art museums and galleries are seen as some of the most important interfaces between art and the public (Duncan 1995; Arvidsson & Werner 2011). Art education in a broad sense is aimed 'to enhance the viewers' ability to see, understand and experience the work of art' (Lindberg 1991: 345f.). To make audiences participate, to move people out of the role of passive observers and into the role of active producers of meaning is one of the hallmarks of twentieth-century art (Bishop 2006). Consequently, the same can be said about the educational practice at museums and galleries. The general public has become an active audience in a meaning making process (Hooper-Greenhill 1994 & 2000). However, it is noteworthy that most contemporary art educational strategies tend to be geared towards children and adolescents, and cognitively able subjects generally, rather than towards people with cognitive impairments. A further progression of art education into a dialogical model of pedagogy is considered one of the most favourable methods for inclusion of a diversified public in the institution of art (see Illeris 2004; Carlgren 2011). In theory, then, the particular participants of the project we have studied, may be thought to benefit from a further progression of art education into a dialogical, participatory model. In this chapter, the question is how participation is carried out in practice. The studied situations contain moments that developed differently than anticipated, showing the complexity of participation in relation to dementia.

---

2. The Malmö Konsthall project started before the Alzheimerfonden and refers to MoMA and to the Norwegian Teknisk museum (see: http://www.konsthall.malmo.se/0.0.i.s/5266, accessed 29 November 2016, and http://www.tekniskmuseum.no/besok-oss/personer-med-demens, accessed 29 November 2016).

Studying participation through participant observations

We conducted a participant observation study during four different sessions of guided tours for people with dementia. Each tour lasted about an hour, but the observation also contained the interval when waiting for the tour to begin, and the tour's closing procedure when photographs of the art works that had been part of the guided tour were distributed to the participants.

Field notes in the form of detailed descriptions, impressions and particular accounts were made directly after the observations by both authors, and comparisons were made between them. Due to the authors' diverse scholarly backgrounds, art history and ethnology respectively, different angles and aspects of the tour were noted. The art educational situations and the way the art educator used different professional tools to create participation and dialogue were highlighted by the art historian, while cultural norms and expected behaviour within this specific context were emphasised by the ethnologist.

Participating in the guided tours involved taking part as a researcher, but also as a visitor or participant in the tour. When the art educator asked the participants to discuss a certain aspect of the present art work, we did as well, with each other or with other participants. The one of us with ethnological background had very little experience of guided art tours, she felt herself drawn into the art educator's storytelling and explanations, adhering sometimes with difficulty to the observational gaze (as well as other senses). The 'participant' in participant observation was certainly accentuated and generated a feeling of 'being there', essential for understanding the situation and context (Frykman & Gilje 2003).

The other participants had been informed of our project when booking the tour, and when the tour started we briefly presented ourselves as researchers studying the art educational pedagogy in this kind of tour. Nevertheless, the blurred line between being a researcher and a participant in the guided tour may have confused the other participants. However, no personal information of the participants was gathered during the observations, and the field notes were made with concern for ethical aspects and the participants' integrity.

Glitches: the humming woman

The field notes from one of our observations at the Art Museum give us several examples of functioning participatory situations but also of glitches. We will describe one instance, which we find significant, as it tells us something about how these encounters may turn out differently from what was intended.

> A woman with dementia and her female friend, both in their seventies, are taking part in the guided tour. They are regular visitors to the museum and its exhibitions. Waiting for the tour to start, the woman with dementia strolls around the entrance area and at the same time she is humming a melody. The humming is tuneful, a distinct but not too loud sound, and it is quite beautiful to listen to. No one seems to pay any attention, though. We are now three people waiting for the tour to start, besides the woman who is humming while she circles around us. We can all hear her, but none of us in the group show any sign of having heard her. We all look away slightly when she comes in sight.
>
> The guided tour is about to start and together with the art educator we take the elevator to the third floor. Throughout the tour, the woman keeps humming her melody. The art educator tells us about the paintings we see, and neither the educator nor any of the visitors seem to pay any attention to the humming. A couple of times the woman stops her humming and speaks a few words about something the art educator has just pointed out about the paintings and the art educator responds appreciatively, trying to prolong the conversation. During the end of the guided tour, the woman stops humming and turns silent. She sits down to rest, apparently very tired.

Apart from obvious cultural norms concerning behaviour, and behaviour in art museums in particular, how can we understand the humming (and its non-recognition) when it comes to ideas and beliefs of participation and dementia? During the guided tour, the woman's humming is ignored and disregarded but her speech is favourably recognized. Why is that? The humming could be seen as an embodied expression, a body-speak if you want, which is a narrative resource. Embodied expressions have, according to the sociologist Jack Katz (1999), several functions. They are simultaneously a bodily doing, an interaction, and a resource for commentary on the social business at hand. As bodily doing, the humming makes the body emerge, and she hums more than she speaks. Perhaps, following Katz (1999), we may guess that if she speaks rather than hums, she will lose touch with her body. Through humming, the woman corporeally realizes her body, her self. The body is situationally transcendent and is thus a secure home for the self (Katz 1999: 273). The habit of humming is perceived as a natural part of the self, easily performed without having to think about how to perform it.

According to Katz, the body is a secure home for the self. But what does that mean? And what is the self, in particular in relation to dementia? This is something that we will analyse further with the help of the concept of selfhood. In the following theoretical discussion, we will look at the concept to find aspects of self, other than strictly cognition and memory, in order to explore participation in relation to dementia and art educational situations.

The neurobiological self

Dementia is perceived as a disease that disintegrates the selfhood of the individual, while the idea of participation is based on full subjectivity (cf. Altermark 2016). When it comes to accounts of selfhood, we argue that participation should be discussed on an existential level. This means the subjectiveness that people gain from each other, seeing and meeting each other, face-to-face (Levinas & Nemo 1985).

Modern psychological science has taken part in shaping an understanding of the individual based on psyche and mind. We as humans see ourselves as bearers of an inner identity, connected to personality and psychological growth (Hansson 2012). The selfhood of the individual is traditionally placed in the mind, but more recently in the brain. Findings in neuroscience have profoundly challenged the Cartesian notions of mind and brain. By studying knowledge practices in biomedicine, the ethnologist Stefan Beck demonstrates how this challenge is articulated:

> While the Cartesian mind was understood as the body's command-centre – the theatre of self-awareness, and the agent of self-identity and self-continuity – new findings disrupted the insular concept of the mind, stressing relations, links and connections to other processes, biological as well as social (Beck 2012: 118).

Beck further analyses how the dualism between brain and mind is currently discussed in neuroscience as well as within the humanities and social sciences. In these scientific traditions, the brain and mind are understood as a truly relational phenomenon (Beck 2012). However, the interpretations of that relational phenomenon, and which part of it that is and should be emphasised, differ. The dialectic view of the brain and mind is furthermore investigated in the book *Neuro*, written by Nikolas

Rose and Joelle M. Abi-Rached (2013). With their background in sociology and history of science, the authors point out that individuals in contemporary Western cultures understand themselves as comprised of a personal interiority and enduring identity to which only they have direct access, and where their true personality, thoughts, feelings and wishes lie. This has not always been the case. Scholars in the humanities and social sciences have long claimed this belief to be a historical and cultural concept. It is an artefact of history, culture, meaning and language. An alternative perspective is to regard human beings as consisting of 'an ever-changing matrix of fleeting thoughts, feelings, and desires' (Rose & Abi-Rached 2013: 202). Rose and Abi-Rached make a point of highlighting a conception of self that was articulated by the anthropologist Marcel Mauss: while the *form* of the self is variable across cultures and epochs, the *sense of self* is universal (p. 218).

In neuroscience, the notion of self tends to be placed in the brain, with an indefinite link to the mind (Vidal 2009). From a neurobiological point of view, mental activity is produced by interconnections and interactions between many relatively distinct components in the brain, each with its own mode of action. Neural organization, then, is not a unitary whole but a multiplicity at molecular level, an assemblage that has come together during the course of evolution. The view of the self has shifted from soul to brain. But, as Rose and Abi-Rached (2013) draw attention to, out of this assemblage of molecules and neurons, 'unless interrupted by pathology or tricked by experiment, a sense of punctual, unified, continuous, autonomous, and self-conscious self always emerges' (p. 214). This notion has psychological roots that still dominate our ways of thinking, speaking and acting and which, according to the authors, are so far unlikely to be effaced easily by neurobiology. Rather, multiple and often contradictory perceptions of self and personhood co-exist.

Nevertheless, in neuroscience, the cultural and historical notion of the self is considered to be a result of evolution. A characteristic of the brain is to generate a self, regardless of historical and cultural context. Therefore, talking about the neurobiological self is not automatically a discussion of consciousness or mind. Rather, the mind appears to be an *effect* of the brain, unable to affect the brain in return. The brain has emerged as the essential organ for the existence of a human self. Rose and Abi-Rached write:

[…] the normal functioning of the human brain has been constituted as crucial for sociality, and all manner of problems with sociality have been linked back to anomalies in neurobiology. This has enabled neurobiological ways of thinking to infuse the analyses of problems of individual and collective human conduct in the many sites and practices that were colonized by the *psy-* disciplines across the twentieth century (Rose & Abi-Rached 2013: 226).

The subject has thus become a 'cerebral subject' (Rose & Abi-Rached 2013: 201). People with reduced cognitive abilities, then, are thus in a sense deprived of their selfhood and humanity; they are seen as lesser of a cerebral subject than others. They are, as the political scientist Niklas Altermark describes it, perceived as lacking logical ability due to anomalies of their neurobiology. They are categorised as 'the Others', compared to those categorised as 'normal' (Altermark 2016).

Another consequence of the cerebral subject is that the brain needs care. Not just any care, but self-care, meaning the overall expectation placed upon individuals to take responsibility for their own lives, especially in relation to health (see for example Alftberg & Hansson 2012). The imperative to take care of the self is something that, according to philosopher Michel Foucault, has become stronger since the nineteenth century. It is connected to the Western project of biomedicine and its striving for control over the sick and disabled body (Foucault 1988 [1984]). The imperative of care of the self has become a part of a medical paradigm where people are expected to take care of their bodies and their health and in this way become good citizens. The science of medicine does not have the full responsibility to cure people; as a citizen, you also have a responsibility to follow the recommendations and guidelines that biomedicine proposes. With the term *biopower*, Foucault has described this change in relation to modern social institutions (Foucault 1995 [1975] & 1990 [1976]). The contemporary demand for self-care and a healthy lifestyle now also includes the brain and involves efforts towards its improvement, ranging from psychopharmaceuticals to brain training and mindfulness.

Self-care and acting on our brains also appears to be more important the older we get. Older people in particular are expected to think about their health and lead a healthy way of life, since ageing is regarded as similar to illness and diseases. Rhetoric arguments concerning older people, uttered by themselves and by others, often involve the mention of brain training rather than pleasure when it comes to activities like solving crosswords or playing bridge (Alftberg 2012). This rhetoric can be interpreted in terms of plasticity. Plasticity is a concept that means that the brain physically changes as a result of learning. It may develop, restructure and adapt throughout life. In contrast to earlier beliefs, in which the brain's connections between neurons were seen as fixed and permanent after entering adulthood, this is now regarded as adaptable, resilient and powerful. Even persons suffering from specific diseases of the brain, which leads to decline in mental abilities, can benefit from brain training in addition to medical treatment (Cohen 2005).

The project 'Meeting with Memories' could be seen as an example, in line with the idea of plasticity and brain training. The purpose of the project is to enhance positive cognitive, social and life quality effects. From a neurobiological perspective, this is achieved by stimulating the brain, mainly visually, through art. To keep on being subjected to cultural artefacts, which triggers sensations and makes different kinds of sense, is to continue being part in an ongoing conversation. The works of art will be looked upon and talked about, thus creating participation between (cerebral) subjects, but the subjects are not equal. Furthermore, as we will show below, this meeting contains cultural norms that affect both the possibilities of participation, and what kind of participation that is desirable.

Embodiment

Rose and Abi-Rached underline that neuroscience, in its search for selfhood, isolates the brain and ignores or denies the fact that brains only have the capacity that they have because they are *embodied*. One cannot understand the brain and its neural processes outside the bodies that they normally inhabit (Rose & Abi-Rached 2013: 230). We briefly described above how the body comes into play in the case of the humming woman, and the significance of embodiment will be discussed further below. First, however, a definition of what embodiment is and how it can be understood is needed.

Embodiment is a concept used in various disciplines with slightly different connotations. According to Rafael Núñez (1999), there exist three understandings of embodiment: trivial, material and full embodiment. The trivial understanding of embodiment only states the obvious, that cognition and the mind are directly related to and made possible by the biological structures and processes that sustain them. Material embodiment is built on this minimal understanding but also adds that the mind is a decentralized phenomenon and takes into account the 'complexity of real-time bodily actions performed by an agent in a real environment' (p. 55). Still, material embodiment is oriented towards more low-level cognitive facts and cannot apply to all events of the mind. Full embodiment is a perspective that shares the previous concepts but further develops the fact that all cognitive processes are sustained by the brain and body. The mind and its activities emerge through specific human bodily grounded processes (Núñez 1999: 55f.).

The link between body and mind is also discussed in the phenomenological tradition. The works of the phenomenological philosopher Maurice Merleau-Ponty (e.g. 2002 [1945]) draw attention to the usually neglected body in philosophical tradition. Merleau-Ponty claims that body and mind are intertwined. The mind is not an isolated entity, separated from the body, but part of the body. The mind could thus be seen as present in every part of the body, not just the head. The point is that the mind is embodied, and it could likewise be said that the body is *enminded* (Ingold 2000: 170). Body and mind are not two separate phenomena, but only two different ways of describing human beings. From that point of view, individuals do not *have* a body, but they *are* their body. This perspective is commonly referred to as *the lived body*. It is impossible to separate the subject – the individual's lived experience – from the body, even though people sometimes may have the impression that the body is not part of themselves and who they are (Merleau-Ponty 2002 [1945]).

So, body and mind are intertwined. But people are also intertwined with their environment. The phenomenological tradition uses the concept of *intentionality* to illustrate how individuals constantly are directed towards something outside themselves. Merleau-Ponty demonstrates how this intentionality always happens through the body. It is as a body humans exist and can relate to the world. This point of departure means that when our body changes, our experience of the world changes as well (Merleau-Ponty 2002 [1945]). This implies a complete turning away from the classical mind-body opposition. The body, its movements, and the relation of this with the environment, play a constitutive role for any perceptual and cognitive act that takes place in the mind (Beck 2012: 131).

Embodiment also enables an alteration; a shift from the dominant attention concerning 'the cognitive side' of the mind towards the significance of emotions and feelings. Emotions can be seen as an important link in the intertwinement of body and mind, since bodily feelings influence and orient human thoughts and behaviour. René Rosfort (2012) claims that emotions are the most embodied of our mental phenomena, and that emotional experience is particularly helpful for understanding the embodied nature of the human mind. During the course of life, 'the body becomes a part of the subject in such a way that the individual is shaped by the body, but the body is also shaped by the subjectivity that it expresses' (p. 98). On the basis of a phenomenological analysis, the author highlights the emotional ambivalence of intimately 'being *my* body and having *a* body' (p. 106); that is the ambivalence of the subjective, lived body and the body as an object, something other than I feel and think myself to be. Rosfort continues:

> This ambivalence of my body as both the centre and the limit of my selfhood surfaces in my emotional life, where the otherness of my body is manifested by the cognitively impenetrable character of my emotional reactions. I cannot penetrate through the biological obscurity at work in my bodily feelings with the rational efforts of my thinking. [...] And still the biological otherness of my body exerts so significant an influence upon the values that inform and orient my existence that I cannot simply ignore the body-speak of my emotional life. [...] My body speaks through feelings that not only shape my thoughts and actions, but also often complicate or entirely bypass reflective engagement with the world (Rosfort 2012: 107).

Embodiment thus sustains selfhood, which is a standpoint that gains importance when we interact with people who are not able to communicate or speak orally. Based on a study of older people with Alzheimer's disease, health researcher Pia Kontos maintains that selfhood, by residing in the pre-reflective body, has two origins. The first is primordial or biological, whereby selfhood emanates from the body's power of natural expression, and manifests in the body's inherent ability to apprehend and convey meaning. Basic bodily movements and social interaction is given as examples of primordial selfhood. The second origin of selfhood is the socio-cultural dimension of the pre-reflective body. This approach necessitates a shift from the primordial to a social level of existence, because socially and culturally acquired behavioural predispositions do not originate in the biological body but are learned. Kontos also argues that the socio-cultural dimension of selfhood includes the dispositions and generative schemes of Bourdieu's concept of *habitus*, which links bodily dispositions to structures of the social world. For instances gestures, which Kontos describes as class distinctions at the corporeal level (Kontos 2004). Accordingly, selfhood cannot be studied as solely a cognitive phenomenon, located in the brain, but selfhood is as much about embodiment and bodily dispositions and gestures, formed by social and cultural processes.

Embodied participation in the museum

We will now return to the situation with the humming woman. As we see it, she takes part in the art educational situation by using her body. The humming is a bodily gesture or expression, showing that she, her self, is participating. Thus, we understand the narrative meaning of her bodily expression, humming, as a way for her to shape her participation but also her social interaction.

Humming is usually interpreted as an introvert action, intimate and private, directed against an inner world. It is something that is commonly done alone or possibly in company of close family members. But at the art museum, the humming could be seen as an extrovert action, directed towards the others. It is a social interaction, a meeting between subjects, creating (or trying to create) participation. The problem is that it is an interaction that does not count within the range of 'normal' communication. When humming, the woman is performing something ordinary and mundane, but it is *out of place*. To appear 'out of place' is characteristic for the construction of disability and non-able bodies. Places are socially produced and organized ways to fit the ideal, healthy and able body, and to block out the non-able body (Kitchin 1998). Places are organized to foster certain behaviours which in turn affects how participation can be carried out. Participation is thus a conditional state, shaped by both physical and social space and interaction. In the rooms of the Art Museum you see other visitors conduct themselves in an orderly manner and you see them looking at you, and looking at art.

The case of the humming and how it could be thought of in terms of taking part in the dialogue triggered by the museum and the paintings, is particularly potent in comparison with the act of singing like another person in the group does at one point. The singing is made normal in the museum in this tour. Not only because the tune is familiar, but also because it is triggered by the content of the painting and interacts with the contextual setting. Singing is encouraged in a way that is recognised, makes sense and is meaningful for the art educators and us other participants. The singing is a shared effort and creates participation that is grounded in beliefs of social inclusion and meetings between people who are active, fully human, subjects. But participation when one of the parties involved is considered to have less of a self, i.e. a disintegrated selfhood, relies on an open approach from the other parties. When the music was played from the art educators' device, a comparison could be made to a concrete example that was present in the room, pointing to other important, shared cultural expressions, while the humming is on a different level, one that we, the rest of the group, do not know and cannot join.

One feature that is thought to create dialogue between individuals, is the use of nametags. It could be argued that this is a method used in educational environments with children and thus perhaps belittling for an adult person. But another way to see this is to compare with the use of nametags in for example a conference venue. It is a way to put a name to a face, to help when addressing people by their name. In the museum, it is used as a tool for setting the stage for dialogue. Other people than the 'person with dementia' are also tagged and are equal partners in dialogue; a dialogue where the actors involved possibly are willing to make their meaning clear in new and different (or same old) ways. Participation is reached when categorisations of selfhood are avoided and when selfhood is recognised as embodied. Nevertheless, open-minded intentions and recognitions may be challenged. The humming woman is one example; another is the incident with the woman who, in the introductory field notes, is described to be wandering off. She might be said to be following the group that seems to be moving more efficiently – or she might be described to have a blurred sense of belonging to the group we were in. In any case this can be seen as a sign that the exclusive setting of being in the museum by ourselves, when it is usually closed for visitors, is an important component, facilitating the visual focus on the paintings and the bodily direction. All of us participants are facing the same way, focusing on the same objects. The disturbing, interfering other group walked hastily by and created another sense of direction. If someone without dementia had gone in that other direction, would they have been called back or would we have trusted their judgement and let them go?

Conclusion: taking part in an art educational dialogue

In the guidelines for the Museum that takes part in the project 'Meeting with Memories', the art educator is constantly reminded that conversation, not a lecture, is the aim. The purpose is not only to impart knowledge but also to encourage participants to engage in discussion. The information given in front of the painting should emphasise the object in front of everyone, and the objects should be limited in number and theme. The questions asked by the art educator should be direct and open. Comparison with other objects should only be made if they are physically close by.

Another point mentioned in the guidelines is the encouragement of all sorts of conversation, as long as it does not disturb other conversations. The conversational aspect is normal behaviour in the museum (but often in low volume). Usually, there are only signs that say 'do not touch', no signs demand silence. The public art museum is often, as in this case, a quite awe-inspiring place with a huge flight of stairs and large rooms that make visitors behave. The museums are designed to house impressive key works of art that a citizen is expected to know of as part of the culture and history of a place, city or nation.

As some of these visitors are people with dementia, their changed and reduced abilities no longer fit the expected conduct at the art museum, that is, to be learning something from the exhibition. But their bodies are not thought to be unfit for conventional social interaction in the museum. Their ability to be affected by the paintings is not believed to have diminished significantly. And they are believed to be able to pick up a conversation sparked by the art educator. At the same time, already categorized as a 'person with dementia', since the guided tour is aimed at this group of people, it may well be that these visitors are presumed by the art educators to differ in behaviour from the ordinary visitor. People with dementia are 'the Other', treated with benevolence, forbearance and acceptance, but their different forms of communication are not always recognized.

We have shown how participation and the intention of art educational dialogue in relation to dementia may be complicated by the idea of the brain as the essential organ for the existence of the human self. The value of cognition and memory means that people with reduced cognitive abilities are attributed a lesser degree of selfhood and humanity. Even though the intent of the art educational situations is to create participation and dialogue, the categorisation and 'otherness' of people with dementia hinder its realisation. To escape the idea that selfhood mainly consists of cognition and memory, we must pay attention to the body and embodied communication. Selfhood is just as much about embodiment and bodily dispositions and gestures, formed by social and cultural processes. Being aware of how participation is affected by cultural norms opens up for an art educational encounter that is equal for all.

References

Alftberg, Åsa 2012: *Vad är det att åldras? En etnologisk studie av åldrande, kropp och materialitet.* Lund: Lund University.

Alftberg, Åsa & Kristofer Hansson 2012: Self-Care Translated into Practice. *Culture Unbound: Journal of Current Cultural Research*, 4: 415–424.

Altermark, Niklas 2016: *After Inclusion: Intellectual Disability as Biopolitics.* Lund: Lund University.

Arvidsson, Kristoffer & Jeff Werner (eds.) 2011: *Konstpedagogik = Art Education.* Skiascope: 4. Göteborg: Göteborgs konstmuseum.

Basting, Anne 2003: Looking Back From Loss: Views of the Self in Alzheimer's Disease. *Journal of Aging Studies*, 17(1): 87–99.

Beck, Stefan 2012: Interlacing the Brain, Contextualizing the Body: Relational Understandings in Social Neuroscience. In: Max Liljefors, Susanne Lundin & Andréa Wiszmeg (eds.) *The Atomized Body: The Cultural Life of Stem Cells, Genes and Neurons.* Lund: Nordic Academic Press.

Bishop, Claire (eds.) 2006: *Participation.* London: Whitechapel.

Carlgren, Maria 2011: Den gemensamma bilden i praktik och retorik. In: Maria Carlgren, Linda Fagerström, Katarina MacLeod & Johanna Rosenqvist (eds.) *Genuspedagogiska gärningar: Subversiv och affirmativ aktion.* Stockholm: Vulkan Förlag.

Caulfield, Sean 2013: The Arts and Dementia: Personal Identity and Self-expression. *Alzheimer's & Dementia: The Journal of the Alzheimer's Association*, 9(4): P650.

Cohen, Gene D. 2005: *The Mature Mind: The Positive Power of the Aging Brain.* New York: Basic Books.

Duncan, Carol 1995: *Civilizing Rituals: Inside Public Art Museums.* London: Routledge.

Foucault, Michel 1988 [1984]: *The History of Sexuality, Vol. 3: The Care of the Self.* New York: Vintage Books.

Foucault, Michel 1990 [1976]: *The History of Sexuality, Vol. 1: An Introduction.* New York: Vintage Books.

Foucault, Michel 1995 [1975]: *Discipline and Punish: The Birth of the Prison.* New York: Random House USA Inc.

Frykman, Jonas & Nils Gilje 2003: Being There: An Introduction. In: Jonas Frykman & Nils Gilje (eds.) *Being There: New Perspectives on Phenomenology and the Analysis of Culture.* Lund: Nordic Academic Press.

Hansson, Kristofer 2012: Crisis and Caring for Inner Selves: Psychiatric Crisis as a Social Classification in Sweden in the 1970s. *Culture Unbound: Journal of Current Cultural Research*, 4: 499–512.

Holmbom Larsen, Axel F., Lennart Minthon & Elisabet Londos 2014: Creativity, Creative Challenges, and Creative Expression, and Their Interaction in Patients Suffering From Neurodegenerative Disease. *Alzheimer's & Dementia: The Journal of the Alzheimer's Association*, 10(4): Poster Presentations: P3-366.

Hooper-Greenhill, Eilean 1994: *The Educational Role of the Museum*. London: Routledge.

Hooper-Greenhill, Eilean 2000: *Museums and the Interpretation of Visual Culture*. London: Routledge.

Illeris, Helene 2004: *Kunstpædagogisk forskning og formidling i Norden 1995–2004: Rapport udarbejdet for Nordisk Akvarellmuseum i forbindelse med projektet kunstpædagogik i Norden*. Skärhamn: Nordiska Akvarellmuseet.

Ingold, Tim 2000: *The Perception of the Environment: Essays in Livelihood, Dwelling and Skill*. London and New York: Routledge.

Katz, Jack 1999: *How Emotions Work*. Chicago: University of Chicago Press.

Kitchin, Rob 1998: 'Out of Place', 'Knowing one's Place': Space, Power and the Exclusion of Disabled People. *Disability & Society*, 13(3): 343–356.

Kontos, Pia C. 2004: Ethnographic Reflections on Selfhood, Embodiment and Alzheimer's Disease. *Ageing & Society*, 24(6): 829–849.

Levinas, Emmanuel & Philippe Nemo 1985: *Ethics and Infinity*. Pittsburgh: Duquesne University Press.

Lindberg, Anna Lena 1991: *Konstpedagogikens dilemma: Historiska rötter och moderna strategier*. Lund: Studentlitteratur.

Merleau-Ponty, Maurice 2002 [1945]: *Phenomenology of Perception*. London: Routledge.

Mittleman, Mary & Cynthia Epstein 2009: Evaluation of Meet Me at MoMA. In: Francesca Rosenberg, Amir Parsa, Laurel Humble & Carrie McGee (eds.) *Meet Me: Making Art Accessible to People with Dementia*. New York: The Museum of Modern Art.

Núñez, Rafael 1999: Could the Future Taste Purple? Reclaiming Mind, Body and Cognition. *Journal of Consciousness Studies*, 6(11–12): 41–60.

Reid, Scott & Elizabeth Hartzell 2013: Art Therapy with a Group of Dementia Caregivers: Exploring Wellbeing Through Social Support and Creative Expression. *Alzheimer's & Dementia: The Journal of the Alzheimer's Association*, 9(4): 485.

Rose, Nikolas & Joelle M. Abi-Rached 2013: *Neuro: The New Brain Sciences and the Management of the Mind*. Princeton: Princeton University Press.

Rosenberg, Francesca, Amir Parsa, Laurel Humble & Carrie McGee (eds.) 2009: *Meet Me: Making Art Accessible to People with Dementia*. New York: The Museum of Modern Art.

Rosenqvist, Johanna & Ellen Suneson 2016: Konst och subjektskapande: Neurodegenerativa nedsättningar och dialogbaserad konstpedagogik. *Socialmedicinsk Tidskrift*, 3: 288–296.

Rosfort, René 2012: Ambivalent Embodiement: Affective Values and Rationality. In: Max Liljefors, Susanne Lundin & Andréa Wiszmeg (eds.) *The Atomized Body: The Cultural Life of Stem Cells, Genes and Neurons*. Lund: Nordic Academic Press.

Rubin, Beatrix P. 2009: Changing Brains: The Emergence of the Field of Adult Neurogenesis. *BioSocieties*, 4(4): 407–424.

Schall, Arthur, Valentina A. Tesky & Johannes Pantel 2015: Art Encounters: A Museum Intervention Study (ARTEMIS) to Promote Emotional Well-Being and Improve Quality of Life in People with Dementia and their Informal Caregivers. *Alzheimer's & Dementia: The Journal of the Alzheimer's Association*, 11(7): 737.

Seigel, Jerrold 2005: *The Idea of the Self.* Cambridge: Cambridge University Press.

Vidal, Fernando 2009: Brainhood, Anthropological Figure of Modernity. *History of the Human Sciences,* 22(1): 5–36.

7. Taking part in clinical trials: The therapeutic ethos of patients and public towards experimental cell transplantations

MARKUS IDVALL

What does it mean to take part in a clinical trial? What expectations do individuals have when they consent to participation in a trial? What do people in general think about clinical research and its particular ways of intervening into participants' lives?

In this chapter, on clinical participation, I will approach these questions by discussing how a certain kind of clinical research – cell transplantations to patients with Parkinson's disease (PD) – is object for an ethical discourse that I call 'therapeutic ethos'. By this latter term I want to scrutinize how lay people relate to a specific form of biomedical science that is associated with uncertainties as well as promises for a particular group in society. Therapeutic ethos implies a mixture of ideas, emotions, attitudes and arguments that together form and sensitize people's ways of relating to biomedical science in general and this clinical research in particular. I am interested in what is experienced as a 'natural' position or standpoint when lay individuals refer to and make choices and judgements about clinical trials.

Overall the chapter is an attempt to identify and establish some of those moral implications that clinical trials in general are imbued with when patients are taking part in these trials. My point of departure is the current discussion on the social and cultural mechanisms for how biomedical research produces not only new treatments but also hopes and imaginations. This discussion has from different disciplinary perspectives problematized how therapeutic promises are formed in different mass media contexts, on the basis of commercial interests as well as through the actions of different stakeholders and organizations (Brown 2003;

Lundin 2004; Novas 2006; Rubin 2008). In my contribution, I want to highlight how this production of hope can be seen as a deeply ethical and embodied engagement that moves individual participants in their active stances towards clinical research (cf. Brown 2003: 7).

I start by describing the emergence of neuron cell transplantations as historical clinical practice. Thereafter I give an account of how the material for the chapter was collected through six separate focus group interviews. The cultural analysis then unfolds through an examination of how people identify with and judge three particular ethical and political issues of the development of neuron cell transplantation to PD patients. First, I focus on the use of placebo or sham surgery and what reactions this procedure creates among lay individuals. Second, I discuss how the enrolment of participants to clinical trials gives rise to different kinds of deliberations on personal risks and possible benefits. Third, I elaborate on how the use – or non-use – of human embryos and foetuses may be linked to different political standpoints. The aim of the chapter is thus to show how people talk with each other about clinical participation and how they, simultaneously, articulate the specific therapeutic ethos that clinical trials are imbued with in biomedical contexts.[1]

The search for a therapy

Parkinson's disease (PD) was identified by the British doctor James Parkinson in 1817. This became the starting-point for many years of search for a therapy for PD. In the early twentieth century, tremor, one of the major Parkinson symptoms, was treated with Belladonna, an extract derived from the plant Deadly Nightshade. In the 1950s a major breakthrough occurred when researchers identified the loss of dopamine, then a newly discovered neurotransmitter, in cases of PD. In the 1960s, Levodopa, a progenitor of dopamine, was introduced as a medication for Parkinson. The new drug made it easier for individuals to live with PD, but also turned out to be connected with side-effects for the individual after some years of medication. Patients who take Levodopa, it was noticed, start to develop different kinds of new symptoms unrelated to

---

1. The text is based on fieldwork conducted within the TRANSEURO programme, and written within a research project called 'The Two Horizons of Research: Cultural Analysis of How Patients and Scientists Face Each Other in Clinical Trials within the Research on Parkinson's Disease', funded by the Swedish Research Council (Vetenskapsrådet) 2014–2016.

the original disease. These symptoms imply unpredictable shifts between periods with good mobility of the body for the patient, and periods with low mobility or even freezing of the body together with the symptoms of the original disease. Furthermore, there are periods when the mobility of the body is exaggerated in the shape of involuntary movements of the body (Hagell 2004: 79; Solimeo 2009: 19–20).

When clinical trials with foetal neuron cell transplants were initiated in the 1980s, hopes were high that this surgical instrument would change the way PD was treated in a more positive direction. From the start in 1987 and over some years, around 400 patients across the world were transplanted on an experimental basis (Barker et al. 2013: 84–85; Petit, Olsson & Brundin 2014: 61). The clinical trials were small in scale and the experimental intervention – the transplantation with the foetal neuron cells – was not in any way disguised or blinded for the enrolled research participants. As chronically ill patients, these individual participants consented to their own participation in the project and were informed about the experimental treatment they were to receive. The projects were based on a so-called open-label approach and the individual participants were followed up after each operation.

Results from these early trials varied a great deal. In some instances, there were clinical improvements for the participants, while in others there were not. Graft survival was however shown clearly in most cases (Dunnett, Björklund & Lindvall 2001: 366). Consequently, in the United States, federal National Institutes of Health (NIH), engaged in these PD issues, decided in the mid-1990s to fund two research projects that were to investigate the effects of foetal neuron cell transplantation in what was rated to be a more controlled and evidence-based manner. With a double-blind approach in the two projects, in contrast to the open-label one in the former trials, 74 PD patients were in total enrolled and randomized into either transplant groups or control groups. At the same time, the scientists were kept ignorant about who belonged to which group among the participating patients. With this model, one hoped to control the transplant outcomes and the research results against possible placebo effects among patients, as well as the phenomenon of investigator bias for how the scientists themselves took action.

When the two projects were reported in the early twenty-first century it was shown that participants in the transplant groups, in contrast to their counterparts in the control groups, had surviving cell transplants in the brain. Moreover, while there were no clear health improvements for any of the groups, the two studies pointed at side-effects – so-called graft-induced involuntary movements or dyskinesias – among the transplanted participants but not among the participants in the control groups who had undergone sham or placebo surgery (Freed et al. 2001; Olanow et al. 2003; Barker et al. 2013: 85).

Here the development of cell transplantation with PD patients could have stopped. Instead, potential cell therapy became even more attractive in the early twenty-first century. The reason was the emergence of human embryonic stem cells as a new possible way to develop a therapy for Parkinson's disease (one among several different chronic diseases). The stem cells made the foetal neuron cells outmoded in a way. At the same time, it was too early to use stem cells in clinical trials with patients. Researchers were still unsure about how the stem cells would react in the brain and there was concern that tumours could develop in the brain as an effect of the cell transplant.

Consequently, when a new multicentre research programme started in Europe in 2010, the goal was to continue the clinical trials with PD patients, but in a way that took the ambiguous situation into consideration. TRANSEURO, the name of the new programme, which stands for 'Neural Transplantation in the Treatment of Patients with Parkinson's Disease', regarded stem cells as the ultimate goal for cell transplantation. Nonetheless, in order to prove the concept, foetal neuron cells were still to be used in the clinical trials that were planned. Foetal neuron cells had in this way become a transitional solution that was to be used in order to learn more about surgical aspects, immunosuppression, patient selection etc. (Evans, Mason & Barker 2012). The initiators of TRANSEURO expected new clinical trials with sham surgery in the future. However, in the meantime they wanted to do 'smaller studies' in Europe (Evans, Mason & Barker 2012: 180).

Focus group conversations on ethical governance

TRANSEURO was a cross-disciplinary programme that included both medical and social sciences. The programme initiated a project on ethical governance, besides the medical studies that were planned, with the aim to map and identify different concerns about neuron cell transplantations among patients and public in England, Germany and Sweden respectively. The project was designed as a focus group study by the social scientist Herbert Gottweis who in different publications had discussed biomedical research from a perspective that linked questions about new forms of governance with aspects of participation (e.g. Gottweis 2008).

In Sweden, the focus group study was conducted by myself together with Professor Susanne Lundin and PhD student Andréa Wiszmeg. We translated the questionnaire from Gottweis for the Swedish context and we recruited focus group participants in accordance with the central project design, which was to interview patients and public in separate focus groups. Eventually we conducted six focus groups with 35 individuals, here presented with anonymized names.

Three of these six focus groups gathered 16 individuals with PD, who were enrolled through the regional patient's association for Parkinson's disease. Each patient group consisted of 5–6 participants and enlisted both women and men (6 women and 10 men all together). The average age of the PD individuals was 61 years; the youngest being 44 and the oldest 78.

The three focus groups comprising persons from the public were represented by 19 individuals. Our first public focus group was formed primarily by persons from the same place of work, namely a university department. The second group chiefly contained neighbours from the same village just outside an urban centre. The third public group was recruited through the regional patient's association and consisted of individuals who were either husband or wife to a person with PD. Each public group consisted of 5–8 participants, with women in the majority in two of the three public groups (in total 12 women and 7 men took part in the public conversations). Their average age was 68 years; the youngest participant representing the public was 33 and the oldest was 86.

The focus group conversations were moderated by either myself or Andréa Wiszmeg (the 'facilitators' in the text below). The conversations were all recorded, and with one exception, they were around two hours in length.[2] Each focus group conversation was transcribed in verbatim and the transcriptions were translated from Swedish into English in order to be a research material that could be used within the transnational TRANSEURO programme.

In Sweden, the project ended in the publication of a report. This report was quite descriptive and accounted for the different themes that

2. One of the public groups, the one joining individuals from the same place of work, was of pilot character and was divided into two sessions during two different evenings. The recording time for this focus group amounted to over four and a half hours.

were treated in the conversations (Idvall, Wiszmeg & Lundin 2013). In this chapter I return to the focus group material to do a more reflective analysis of the focus group discussions.

Bodily harm, sacrifices and therapeutic returns

Sham, or placebo, surgery was as mentioned above a crucial instrument in the two surgical studies on foetal neuron cell transplants that were funded by the National Institutes of Health (NIH) in the mid-1990s. In this controlled and evidence-based research, the information about who received the cell transplants and who did not was unknown to both the participating patients and the scientists, with the exception, of course, of the operating surgeon. Thus, all participants in the two projects, these seventy-four individuals, received brain surgery in some form. Individuals belonging to the transplant groups were implanted with foetal neuron cells that were expected to grow in the brain and start producing dopamine, whereas the individuals allotted to the control groups received small drilled holes in the skull but no implantation into the brain itself (Freed et al. 2001; Olanow et al. 2003).

In the ethical debate that arose, supporters of sham surgery stressed the scientific or methodological rationale for the use of sham surgery when doing foetal neuron cell transplants (Miller 2003: 41). It was seen as a possible way to avoid 'false positive trial results' due to placebo effects or investigator bias (Albin 2002: 322). Risks were in this case perceived as something that was 'justified' and 'reasonable' and that could be reduced (Freeman et al. 1999: 988, 990; Miller 2003: 43). Critics of sham surgery, in contrast, focused on 'ethical concerns' and emphasized the researchers' responsibility to reduce risks for research participants (Macklin 1999: 993). In the critics' view, there was also a risk for 'therapeutic misconception' among research participants. These may have consented to participation in the project on the basis of what was seen as unrealistic expectations about being cured (Macklin 1999: 994).

When we introduced the issue of sham surgery in the focus groups it gave rise to animated discussions. Our starting-point was the straight question whether sham surgery should be accepted in clinical trials at all. We soon realized that this method will never be uncontroversial among lay individuals. In one of the public groups, sham surgery was debated in the following manner:

MATTIAS: It's not just, not doing anything, taking a sugar pill, but you are *hurting* the patient, by drilling a hole into the head.

AGNETA: Yes, because you go *in* and do something!

JONAS: Okay, so you mean, physically, that you …

FACILITATOR: It's an active procedure, so that you can …

JONAS: Okay, okay, I'm not so sensitive about that.

[Mattias laughs.]

When discussing sham surgery, the participants focused on the human body and framed their arguments in terms of the bodily harm for the research patients in the control group. In one patient focus group, the facilitator, hoping to broaden the discussion, tried in vain to make the participants think anything positively about sham surgery.

FACILITATOR: I think that this kind of sham surgery, it's ethically sensitive, but would it be acceptable for this kind of trial?

LARS: No.

ANNIKA: I really think it's tricky.

MALIN: Well … if you take a pill, a sugar pill, that's one thing, but to have an operation … An operation is a risk regardless of what you do. There's always a risk. That you should do it without having anything done! To open up just anyhow! No, I think it seems really strange, I have to say.

[Murmuring in the background.]

ANNIKA: There is so much at stake.

By questioning how surgeons may 'open up' the body or the skull 'without having anything done', Malin summarized the criticism from her focus group against sham surgery. This group of five PD individuals, four women and one man, all agreed that sham surgery must be seen as a risky business that should preferably be avoided. Sham surgery appeared as too invasive and unsafe, not only in itself but also in comparison with how placebo is used within pharmaceutical studies. This comparison was also present in another patient conversation. Here, Gunnar, a man in his mid-sixties, reasoned in this manner:

Of course it's always, if you … Cutting in people is a risk. You expose … I mean I've done … I've participated in trying out a medicine that was supposed to reduce the [involuntary] mobility or dyskinesia, and then we didn't know who got sugar pills and who got ... but that's not as drastic as cutting in people.

There were some different opinions, however. In the public group quoted at the beginning of this section, Mattias and Jonas – the two men in the group – took an opposite position. Mattias was firmly against sham surgery: this kind of intervention would, in his eyes, hurt a patient who would not gain anything from the surgery. Jonas was not enthusiastic about the procedure, but obviously also wanted to discuss the potentialities of this kind of operation, maybe in relation to the pharmaceutical alternative. In the exchange of views that continued for several minutes and involved everyone in the group, Mattias eventually declared that, from his point of view, a clear ethical boundary or line existed when it came to sham surgery:

MATTIAS: I mean, there's a line in this core-trial where you hurt the patient. [...] It's not that I'm giving air to a concern or that it's scary, or that maybe I'm hoping for something. On the contrary, there's a line that I simply draw and that from that line it's ...

FACILITATOR: So from this point, you find it inconceivable, therefore you think that what you feel about it doesn't matter, because, or ... it's not negotiable?

MATTIAS: Right! It's not anything that I'm interested in being part of, you know, for an empirical mapping of [...] different people's kind of feelings on a sliding scale, which already from the outset was a slippery slope. [– – –] I will not enter that slippery slope, so to speak.

After having listened to these reflections of Mattias and the facilitator, Jonas explained that he wanted to 'make a certain reservation':

JONAS: There is this, if it's good, there may be new stuff ... So, yes. So it's this whole thing that if one would assume that ... that there are ... How shall we put it ... There are ...

MATTIAS: You are hoping for the best?

JONAS: No. Not that way. I'm probably rather thinking that ... literally speaking, that is, that if we make experiments and this is good for humanity, so to speak ...

MATTIAS: Even if we make a sacrifice?

JONAS: *Despite* the sacrifice, yes.

MATTIAS: Half of humanity?

JONAS: But new elements may emerge, so to speak. In thirty years … yes …

AGNETA: But would you personally want to be the placebo patient? Would you, because that's a premise, that you can imagine that. […] Could you imagine that [being a placebo patient]?

JONAS: Pfff, yeah somewhere, maybe someone would, perhaps somehow anyway.

Another participant, Agneta, who during the discussion had seemed to be close to Mattias's point of view, thus asked Jonas whether he himself was prepared to be that placebo patient in a procedure that he apparently was prepared to defend. This question made Jonas become momentarily a bit taken aback. Still, he insisted that the question should be discussed and that, to him, sham surgery was not 'one hundred percent black as night'.

In this focus group with public participants, *sacrifice* thus appeared as an appropriate metaphor for what sham surgery, in the worst case, may involve for those who were operated on but did not receive the active cell transplant. This metaphor was suggested by Mattias when he raised the straight question – '[e]ven if we make a sacrifice?' – and Jonas, his antagonist, confirmed by emphasizing the ambivalent situation: '*[d]espite* the sacrifice, yes'.

Clinical trials with sham surgery thus seem to involve sacrificing victims in the eyes of lay persons. According to our focus groups, these victims are primarily the control group participants, who not only consent to be deceived into receiving a fake intervention of the skull even though they want something else, namely an effective treatment, but who also eventually receive the faked procedure with material, embodied consequences.

For this victimization to be avoided within sham surgery, the focus group participants argued, there should therefore be some kind of therapeutic return benefitting the trial patients in the later steps of a study or subsequently. One patient focus group talked about the possibility of organizing sham projects in a better manner. They argued that if operations were successful in a sham project, those trial participants who received surgery but no active substance – the placebo group – should, as one group participant expressed it, be promised to be 'ahead in order of priority for successful surgery' after the trials were finished.

Therapeutic return that counters victimization can also come in the shape of information. Straight information in advance may make sham surgery more acceptable as an experience. In one of the patient focus groups, Åke referred to the importance of information when this topic was touched upon. In this group, Kerstin clearly opposed sham surgery by exclaiming 'madness' when she first heard the facilitator's question about it. In the discussion, it turned out that Kerstin had the other people in the group on her side. Still, Åke did not want to reject the usefulness of sham surgery. In his view, it was something that one can 'take', when being a research subject. He himself was prepared to deliberate on his own possible participation in, or dismissal of, this kind of project if he got proper information about why sham surgery must be used:

> ÅKE: I want to know how they base their decision to do placebo or not.
>
> FACILITATOR: Yes.
>
> ÅKE: I would've asked that question. Reasonable answer to that …
>
> FACILITATOR: Mm.
>
> ÅKE: So then I'll [inaudible] say yes or no.

To sum up so far, a therapeutic ethos appears in how the focus groups reasoned about sham surgery, in the sense that the highly experimental method in question mostly was approached by the individuals, not as a neutral instrument which must be applied for scientific purposes, but as a procedure that is quite controversial due to its limitations regarding therapeutic benefits in the present. Most group participants were indeed either opposed to, or hesitant about, this specific technique due to its presumed harmful invasiveness. But for some, the resistance towards sham surgery was mitigated. Therapeutic return, either in the shape of actual surgery executed afterwards or in the shape of patient information, was seen as something that could counter or take away the feeling that something was wrong with a surgical intervention that intentionally harmed individuals physically without giving them any clear or expected benefit.

Synchronizing the self with science

This discussion that the focus groups had on sham surgery raises questions about why people would participate in clinical trials that may not be of immediate benefit to them. In the two above-mentioned NIH studies that were conducted in the United States in the mid-1990s, more than seventy individuals gave their informed consent to participate in the planned sham surgery protocols. Of these, around 30 individuals were randomly assigned to a placebo group and were eventually transplanted in a fake way (Freed et al. 2001; Olanow et al. 2003). In the scientific literature, we learn very little about the friction, if any, that this possibly caused in the contacts with the patients. One of the articles briefly mentions that 'one [patient] withdrew consent after 1 month' (Olanow et al. 2003: 405). Neither do we get any exact information about how many patients, if any, who rejected taking part in the procedure after being informed about the clinical trials. Thus, we do not know how these individuals with Parkinson, together with their families, reasoned when confronted with the recruitment initiatives of the NIH trials in the 1990s. But the focus group participants' reactions to sham surgery and, in extension, to questions about participating in clinical trials, may give us the beginning of an answer.

The experience of agreeing or not agreeing to participation in a clinical trial was not new or unknown in the patient focus groups. It was well represented in all three groups. There were individuals who had taken part, not only in pharmaceutical trials but also in brain surgery trials, either with implantation of human growth factor or with Deep Brain Stimulation (DBS). Of these two interventions, the first is on an experimental level and the second is a type of specialist treatment for involuntary movements of the body (dyskinesias).[3] Even more significant for the discussion was the fact that individuals in two of the three patient focus groups had accepted to participate in the clinical trials of transplantation with foetal neuron cells that TRANSEURO were planning. Based on what these individuals with trial experience expressed in the conversations, one can distinguish what I would call a form of *synchronizing the self with science*. By this I mean a process whereby lay individuals, in temporal terms, relate to the possibilities of participation in clinical

---

3. The experiments with the human growth factor, as described in the participant information in our focus group study, are about developing a system that can stimulate the growth of new nerve cells in the brain. A certain protein, which exists naturally in the body, is procured from platelets and supplied to the brain. The goal is reached when the patient starts to produce her/his own dopamine. Deep Brain Stimulation (DBS) surgery involves the implantation of a small apparatus with multiple electrodes into the brain. A neurostimulator, placed underneath the skin next to the collar bone, communicates through a wire with the electrodes in the brain; the patient can control her/his involuntary movements through the pulses that it generates.

trials and hence match or dis-match their own individual movement into the disease with the parallel institutional movement of the scientific practices.[4]

In one of the patient groups, traces of this synchronizing process appear after the discussion had run for half an hour. The facilitator, myself on this occasion, asked if anyone in the group had any experience of transplantation with foetal neuron cells. On this direct request one of the men in the group, Peter, decided to share his experience of having accepted to participate in such trials. He was now 'included in such a group' that will be assessed for being eligible to take part in clinical trials of transplantation with foetal neuron cells. Before this contribution to the discussion, Peter had been quite silent. Only two minor contributions stood out. In his first statement he had declared, in agreement with the rest of the group, that he supported research on the foetus and that this kind of research was a way to 'reach the core of the disease somehow'. In this first contribution, he also briefly mentioned that he was taking part 'in these kinds of projects' himself. He could identify with the disappointment that one of the other focus group participants had expressed about how Parkinson research had been hindered earlier by political developments in the United States (below I will return to this political question). Further on, in his second contribution, Peter explained that he understood 'the ethical problems' with foetal neuron cells, but added: 'I still feel that there is so much to gain here in this, so that overshadows it for me.'

Peter was in his mid-fifties. At least to me, a cultural scientist with no medical schooling, he showed few, if any, symptoms of Parkinson on this early afternoon that we met. He was not prominent in the conversations in general, but he was well articulated in those parts of the discussion when he joined in. With his academic background, he was still working full-time. In contrast to the other focus group participants he eventually left the room in a rush in order to get back to work. When he told us about his enrolment in the TRANSEURO project, he suddenly became the centre of everybody's attention. He explained how he had reasoned before accepting participation. Reflecting on the transcript of this patient focus group, I now interpret this explanation as an account of how his synchronizing of himself with science took shape when he

---

4. For a theoretical discussion of this concept of synchronizing the self with science, see Idvall (2017).

was approached by/approaching the clinical trials in question. Of course, he had deliberated on what effects and possible side-effects the trial participation would have for him. He had tried to estimate the progress of his disease in relation to the whereabouts of the scientific progress within the research field of foetal neuron cell transplants. 'I guess', Peter said, 'I've been thinking a lot about this: Does it [the clinical trial] affect my disease? How far along is it [research in relation to anticipated results]? Is it really early in the trial stage?'

After allowing the other participants to converse for a minute or two, Peter re-entered the discussion and tried to summarize the agony he had felt before consenting to the cell transplant project:

> But I think, like you say … to go into this, I have just now had those feelings: Is it worth it to accept? Or does one have to feel … or have advanced further with the disease? … It should, sort of, be worth a try. And like this, I've been struggling with these questions, since I'm in the early stages of the disease. Well, early, sort of. After all, I've had the diagnosis for five years now and had the symptoms maybe a few years more. But it's easier for one who's further into the disease … who's having it harder to deal with everyday life. Then in some way it is like you're saying: It can't get much worse! Maybe it can, but it's significantly … so, along those lines.

During Peter's long reflection, the other participants kept humming affirmatively. When Peter finally stopped, another man in the focus group gave a sort of conclusion to what they all had listened to: 'But it's still hard to know where to put the limit', meaning in what situations and phases of one's life to say yes or no to trial participation. Peter apparently did not want to expand on the question about the limit, and only replied a short yes to the comment. However, questions about when the limit is reached for the degeneration of the own body in parallel with the unfolding of research are central for potential participants when they deliberate on synchronizing their selves with clinical trials and consider whether or not they should take part. These limits or thresholds often define the whole situation for those who weigh the pros and cons of participating in a clinical project.

Synchronizing the self with science is a theme that also appears in one of the other patient focus groups. Here Annika, a woman in her mid-forties, was active in the conversations from the beginning. Being among persons whom she knew well, she seemed relaxed and did not hesitate to joke about things. Together with her friend Malin, she commented on what the two women saw as a major change that has occurred over the last ten years: today, you can live with Parkinson's disease almost as long as a normal life span. She also confirmed her friend's statement that patients often have to wait too long and become too ill before they get treatment. She furthermore agreed with her focus group as a whole that the practice of using aborted foetuses for research should be allowed, since 'society permits abortion'.

Annika's statement about her decision to try to become enrolled in clinical trials, which indirectly disclosed her experience of synchronizing the self with science, came in the middle of a long passage that started with Malin describing her personal stress about the medications that she was taking. These drugs had worked fine in the beginning, but now Malin had started to experience that they were losing their effects. She needed to take more and more medicine in order to feel well. She perceived that she was becoming enslaved by the medicine while, as she expressed, the clock was ticking. 'I will not', Malin concluded, 'be able to take a whole lot more, because there is a limit.' Malin had therefore started to consider trying to get a DBS operation, but she felt nervous about its possible side-effects. If her voice, for example, was to become even weaker than it already was with Parkinson's disease, she felt that such a surgical treatment was not worth trying. Annika and Malin now engaged in a dialogue which ended in the conclusion that the medicines that the two of them were so dependent upon at present were really not the solution in the long run. At this juncture, and after reaching this mutual conclusion, Annika declared that she was now actually trying to take part in 'this TRANSEURO project'. She explained that 'if I get the chance I will take it, because I feel that the alternatives are not so nice'. In the light of her friend's reflections on the limits of medication and the possibility of a Deep Brain Stimulation operation, it was clear that Annika saw her acceptance of clinical participation as a movement in harmony with how she experienced her own degenerating embodied condition. She set it in relation to a development within science that she found promising in many ways, but of course, she was uncertain about its progress.

The contributions by the two potential TRANSEURO participants – Peter and Annika – initiated further discussions in their respective focus group. These concerned questions of taking part in trials and covered both the hardship and the potential success that such trials may offer. In Peter's focus group, a man told of a person who took part in a research project on the human growth factor (for this technique, see note 3 above). This person was forced to do 'very extensive tests' and 'spare a whole year, lying [in bed] at Karolinska [Institute]'. Hearing this, another group participant commented by saying: 'you've got to be really sick if you're going to do that [...]. You probably wait as long as you can, and then when you're that bad you don't give a shit'. In Annika's focus group, her decision to participate in clinical trials was approved of by her Parkinson friends. One of them concluded that 'you're on your own [...] [and] kind of taking all possibilities that you're offered'.

Synchronizing one's life and one's Parkinson with what one experiences as the progress of clinical science is thus a sort of temporal practice. Its cultural characteristics, involving different kinds of personal and embodied engagements with those institutional processes that science is constituted of, are recognizable for groups with other chronic illnesses, too. The synchronizing practices of the PD individuals can be likened to how sociologist Kathy Charmaz (1991) has described the life of chronically ill people. In her analysis of how people experience chronic illness and cope with problems in everyday life, she shows how living with a chronic illness is a matter of experiencing and structuring *time* in specific ways. The chronically ill patient's 'self', Charmaz (1991: 4) argues, lives his or her life 'in time'. Hereby, in relation to orientations and perceptions, the chronically ill person frames, organizes and uses time in strategic manners. Charmaz does not specifically discuss how chronically ill people relate to biomedical science, but I think that her theory of 'the chronically ill self in time' could be developed, using her discussions on how one as chronically ill lives in an 'intense present', both in relation to a possibly 'improved future', but also to what could be seen as a 'dreaded future' (1991: 250–253).

What is striking is that the synchronizing of the self with science appeared as a strategy not only in the patient focus groups but in the public ones, too. Here, the participants' thinking naturally did not have the concrete, embodied features that it had in the patients' discussions. Instead it showed more of an empathy-based way of understanding how one, as a chronically ill individual, would reason in one's contact with clinical science. There was often a focus upon the chronological age of the potential participant and on how ill – how 'desperate' or even 'close to death' – he or she would have to be in relation to what the research project could offer.

GUNNEL: Yes, I would never have dared to have an operation if I had been …

FACILITATOR: No, no.

MARIANNE: But it depends a little bit on what stage you are at. You don't think you get so much worse, and then maybe you do it anyway, if they now can do one of those operations so you go along with it.

FACILITATOR: Yes.

MARIANNE: Because you can, the hope is that you might become better and someone of course has to try it to see … I think that for the research to go forward, someone has to try it too.

BARBRO: If you are fifty years old, for example, and you try and if it is still not successful, then it is not very …

MARIANNE: No, but if you are really, really bad, and if you are sixty-five years old and you think that I no longer want to live with this, I will do what I can, try it on me and do it. I think, I would do it, if I were, yeah.

The participants in this discussion (apart from the facilitator) were all non-patients in their seventies and eighties. They wanted to reach a consensus on when in life the critical 'stage' is situated when one would accept clinical participation. This stage apparently comes when one's chronological age is juxtaposed to one's perception of whether one is about to get 'better' or 'worse' in the present situation. To accept participation in a trial appears as difficult work in informational terms. To a large extent it seems to be an act of balancing between different kinds of multi-facetted information originating from social interactions with doctors, nurses and scientists as well as with other patients and from one's own body.

In one of the patient focus groups, Kurt articulated this difficult informational work of coordinating the progress of one's self and one's disease with the development of science. He expressed his disappointment when taking part in a specific research project, which eventually had led to a feeling that his life had lost in quality since the results from the project were quite modest compared to what he had expected when he enrolled as a participant.

> In this first attempt, which was one and a half years ago, I got a medium sized dose and I received medicine for 14 days which was pumped into the brain and the result was that I could feel smells better. That was the only result that was noticeable.

Now in his mid-sixties, Kurt had been ill in Parkinson's disease for more than ten years. He had growing problems with dyskinesias and hoped for something more than a better sense of smell, an effect that even faded away after some months. He continued:

> If I had known that I would be walking around with this pump [an obligatory device in the research project, which is connected to the body] for two years, I don't think I would have gotten into it. So that's no fun. What they've said is that they're going to move on with the next project, but they're amazingly slow.

Kurt told us that he wanted to continue in the project. But he was annoyed about not getting any information from those managing the trial: 'What are they doing? What are they thinking? What do they want?' In line with his questioning of the researchers, Kurt referred to a friend who had also taken part in the research project in question, but who after some time as a trial participant chose a different direction. The friend, Kurt explained, had finally said: 'No, I'm not waiting any longer! Give me a DBS right now!' And Kurt continued: 'He's very pleased with his choice. [– – –] He quit a year ago and had a DBS surgery instead.'

But withdrawing from a research project can be as difficult for a trial participant as the initial agreeing to participate. The patient who leaves may be seen as the one who 'ruins everything', as one of Kurt's focus group partners expressed it. A research patient may have a feeling of obligation towards the researcher, who might be one's doctor too, to stay on in the clinical trial, to finish what has been started and to be a part of the final results, since this is what one's embodied participation in some sense was meant to guarantee originally.

To sum up these discussions, it seems that synchronizing the self with science is not only based on aligning one's self-interest with how scientific progress is perceived. It also builds on a particular ethos of how obligations and perceptions of responsibility should be negotiated in particular contexts. We are now closer to understanding the therapeutic ethos presented in the beginning of this chapter. I have linked its features to how the self is enacted in relation to a complex of significant others in the shape of doctors, nurses and scientists as well as family. To conclude the discussion, I will add a third dimension of this therapeutic ethos; it concerns how people judge and identify with foetal neuron cell transplants in the context of clinical trials. Simultaneously, this moreover moves the phenomenon into a political context.

Cell transplant controversies

I will now turn to how the focus groups discussed the use of human embryos and foetuses as a source for cell transplantation in clinical trials. This has a political dimension as well as a scientific one. Scientifically, this cell source has been defined over some decades. In the two NIH funded studies that took place in the 1990s, the source material was neuron cells from six to nine week-old aborted human foetuses (Olanow et al. 2003: 404; Freed et al. 2001: 711). The foetal neuron cells had at the end of the 1980s been selected to replace cells from adrenal glands as the primary research material for cell transplantations to PD patients (Idvall 2017). However, and as mentioned above, only a few years later, in the early twenty-first century, the use of foetal neuron cells was challenged by a new contender in the field: the human embryonic stem cells. Today, scientists expect such stem cells to become the primary research material and eventually also provide the solution that may heal people from Parkinson. Thus, human embryonic stem cells show great potential in the long run, but they are at the moment associated with various problems, including a cancer risk, which need to be solved before patients can be exposed to this technique (Devolder 2015: 11).

Politically, and in parallel with the scientific development of this source material, resistance has developed against the scientific use of foetuses and embryos, especially in the United States where the sham surgery took place in trials with foetal neuron transplants (Gottweis, Salter & Waldby 2009). The US Supreme Court legalized the right to abortion in 1973, but an anti-abortion movement has since then fought against the legal right of women to have an abortion, and therefore indirectly against the scientific use of aborted foetuses. The resistance against abortion turned into a parallel opposition against research on the human embryo when cloning and human embryonic stem cell research appeared as the new biotechnologies of the 1990s. Different views collided on how to define what human life is. On the one hand, there was the pro-life movement that battled for what it considered to be the rights of the human embryo. On the other hand, biotech companies, risk capitalists, individual researchers and patients' organizations stressed the importance of stem cell research for finding new treatments and cures. In the global setting, research embryos were experienced as controversial and their use was fully accepted only in a handful of countries (Gottweis, Salter & Waldby 2009: 144). In the United States, the use of research embryos was regulated on a federal level by Presidential decisions. The Republican president George W. Bush declared on 9 August 2001 that no production of new stem cell lines funded by federal money was to be allowed from the current date, but stem cell lines that already existed were to be permitted even after that date (Wertz 2002: 675; Gottweis, Salter & Waldby 2009: 101). The private sector which had invested a great deal in stem cell research was not included in this decision; it was 'largely untouched' by these federal regulations (Gottweis, Salter & Waldby 2009: 102). Eventually in 2009, Democrat President Barack Obama removed Bush's cut-off date and made it possible also for federally funded researchers to produce new embryonic stem cell lines (Devolder 2015: 101–103).

When we introduced the focus groups to the topic of the two source alternatives – foetal neuron cells and embryonic stem cells respectively – the discussions became largely focused on the first one. Neuron cells from aborted foetuses seemed to be a more concrete topic for the focus groups; something that everyone could more easily relate to within the frames of the therapeutically orientated questionnaire that structured the conversations. However, there was a difference between the public and the patient groups considering *how* the group participants talked about the foetal neuron cells. In the public groups, the foetal neuron cells were connected with dilemmas that primarily concerned the problem of abortion as a source of material for science. In one of the public groups the participants discussed how the donors were to be approached in the abortion situation. They considered that the 'parents' of the foetus must give their informed consent about such a usage and they debated about the specific terms of how this information should be delivered. Should the donors be informed about how the foetus was to be specifically used? Or would it be enough with general information about the scientific use of the foetus? One focus group also problematized abortion as a source of research material by linking it to risks of the foetus being commodified in an undesired way.

> AGNETA: As I understand it, where they get the foetuses, that's the abortion clinics, but is that something that could be developed into merchandise?

> FACILITATOR: I think it could, yes.

> JONAS: I think I had a little note about that.

> HELENA: I'm sure it is, in a grey economy.

AGNETA: So now we're talking about, like, might there be countries where foetuses are produced because of profits, so to say.

MATTIAS: That's imaginable.

AGNETA: How ethical is that!

GERTRUD: Well of course! [Laughs.]

FACILITATOR: No, but this is problematic, and it's the kind of thing we …

AGNETA: It's not good.

Given these reservations, using foetal neuron cells was seen as a temporary stage in the development of research. Embryonic stem cell transplantation, which do not rely on abortions, but on discarded embryos from in-vitro fertilizations, was experienced as a more promising alternative in the long run by this public group. However, one of the other public focus groups, the one whose participants were spouses of PD individuals, argued that abortion as a source for research material should not be questioned. If the decision had been taken to perform an abortion, then it should be no problem for the donor of the foetus to allow it to benefit somebody else, in this case, scientists involved in research or, in the longer perspective, people benefitting from whatever research may achieve.

ERIK: The thing is that it's an aborted foetus, and it's like it's already aborted or lost or whatever you call it.

MATS: Yes, exactly.

ERIK: It's already …

MATS: It's a choice that someone has made themselves, you can hope.

FACILITATOR: Mm.

ERIK: Yes, really.

MATS: And then, I don't think … if we can make another person happier, so to say, then I think it's completely okay.

This particular focus group also discussed the ethics of abortion in terms of how differently an aborted foetus can be valued: on the one hand, as something that can be memorialized and buried at a funeral or, on the other hand, as something that is rejected and thrown into a sink.

Thus, the public groups tended to scrutinize the situation of abortion, irrespective of whether one was for or against the use of foetal neuron cells as research material. In contrast, the patient groups paid attention to the general political context of the specific cells. In one of the patient focus groups, Åke started the conversation on the possibilities of receiving a foetal neuron cell transplant by mentioning the name of a well-known political character.

ÅKE: I think Bush should have had a red card a long time ago.

FACILITATOR: Eh?

ÅKE: They stopped research all together …

FACILITATOR: Yeah, that's right.

ÅKE: On foetuses. They [who stopped research] were opposed to making abortions.

FACILITATOR: Yes.

In Åke's eyes, former president George W. Bush was a representative of the conservative resistance against the research on human foetuses that PD scientists depend upon for their work. Bush and his supporters ('they' in the quote above) thereby also became a barrier for Parkinson individuals' access to research and clinical trials and, in the longer perspective, to possible treatments and future personal health. What Åke was thinking of when mentioning the name of the earlier president was probably his aforementioned decision in August 2001 to limit federal funding of research on stem cell lines. Even if it is not clear whether Bush's decision was harmful for the development of stem cell research or not, in the focus groups his name triggered a discussion about barriers for the progress of research.[5] Göran associated to what he had experienced some years before President Bush's decision on stem cells in 2001:

GÖRAN: Then, in the late 90's, when I was a part of this research project, and Bush came to power, I was bloody disappointed, because I was kind of lulled into an expectation that there was a possibility to track down maybe not a cure but the disease within a period of five or six years. This was -97, -98.

---

5. Herbert Gottweis, Brian Salter and Catherine Waldby (2009: 101–102) argue that it would be wrong to understand the decision by President Bush on 9 August 2001 'as a reversal of the Clinton administration approach'. On the contrary, the decision by Bush 'was in many ways consistent with the previous policy'. The decision even 'strengthened the position' of some of the private actors in the stem cell research in the US.

KERSTIN: But does that mean that researchers were prohibited to do this [research] in the US?

GÖRAN: I don't know, I haven't a clue.

ÅKE: It was him [Bush] who was in a leading position.

FACILITATOR: Yes.

Göran was mistaken about Bush coming to power in the late 1990s since at that time Bill Clinton was still in office. But he was in a way correct, in terms of pointing to the general atmosphere for research on embryos in the United States in these years just before the millennium shift. This specific kind of research was debated and also entangled in a struggle with the pro-life movement (Wertz 2002). For Göran, Åke and the other participants in this patient focus group, the long tradition of resistance against research on the human embryo in the United States represented a kind of political barrier for free science, and indirectly for the possibility for individuals to get a potential therapeutic cure in the future. One participant in the group summed up the situation:

HENRIK: There are a lot of people, especially in the US, who are opposed to abortions, and they … in some way you have to accept their opinions as well.

FACILITATOR: Mm.

HENRIK: But … on the whole this is, as I see it, a form of organ transplant, just as we transplant heart and kidney, to use foetus, aborted foetus, it's nothing special. So, the decision that this human being is not going to live anymore is already decided some time in advance by the parent.

Thus, the patient focus groups tended to highlight politically and ethically charged questions about access to possible research material as well as issues of inclusion in contexts that may lead to scientific breakthroughs. Relating the use of the human foetus to the relatively normal and well-established practice of organ transplantation, as is done in the quote above, was one way to take a stand for research on the foetus in opposition to abortion opponents and the pro-life movement; and at the same time to include oneself in this particular research endeavour. To Henrik and others in this patient focus group, abortion was not anything that they or the scientists had a part in. The decision to end the life of the foetus was, in their eyes, taken by somebody else in a situation that did not involve PD individuals or issues. Therefore, these focus group participants saw no obstacles against using the foetus for cell transplant purposes and research.

Thus, two slightly different, but overlapping, ways of relating to the embryonic stem cell alternative appeared in the discussions. In the public focus groups, the alternative of the embryonic stem cells was viewed by many as what science should target in the long run as a means of avoiding the problematic donor question. The embryonic stem cells were seen as much better than foetal neuron cells, since using them would mean that one did not need to contemplate the difficult question of the aborted foetus. The patient focus groups, on the other hand, were slightly hesitant about the embryonic stem cells as a better alternative than the foetal neuron cells, because the former were not considered to be as accessible as the latter. Thus, they also wanted researchers to continue work on the more accessible and realistic alternative of the foetal neuron cells. This difference between the patients' and the public's views on stem cells and foetal neuron cells has been explained by Andréa Wiszmeg (2012: 75–81) as based on a distinction between the pragmatism of patients and the effects of a more reflexive approach of non-affected individuals.

Conclusion: the therapeutic ethos of clinical participation

In the introduction to this chapter I asked what it means to participate in a clinical trial. What hopes and fantasies do people have when they reflect on taking part in clinical trials? I have discussed the phenomenon of clinical participation by examining how people talk together about a certain example of biomedical research: the clinical science of cell transplantations to patients with Parkinson's disease. I was interested in a certain ethical discourse of health promises among the lay individuals, which I call a 'therapeutic ethos'. This I translate as a morally charged and embodied engagement with different aspects of clinical trials (cf. Brown 2003: 7). Based on a focus group material, I have studied how the specific therapeutic ethos was articulated in the conversations in groups involving either patients or the public. In this latter case, non-patients were confronted with the specific questions concerning treatment alternatives and cell transplant trials for PD individuals.

Sham, or placebo, surgery was the starting-point for my analysis of the therapeutic ethos of clinical participation. Through its association with bodily harm, sham surgery is an effective frame for clarifying the importance of therapeutic return in relation to the risks of making participants into victims. I then discussed the reflective and deliberating practice of synchronizing the self with science, which is central for how lay individuals relate to the possibilities of clinical participation. This section ends up in the mapping of a certain temporal practice that balances how the individual perceives the progress of the scientific project in relation to her or his own progress into the illness. In the last section, the fact that the sham surgery trials were dependent on the use of human foetuses was the rationale for a discussion on how clinical participation also can have a political aspect. Beyond questions about sham surgery, risks and benefits, a political landscape emerges on the basis of the complexities of the different transplant sources that exist in the shape of human foetuses and embryos. In their wishes to have access to cell transplants that may heal them, patients with Parkinson stand against representatives of the anti-abortion and pro-life movement.

The discussions of the focus groups thus show how the therapeutic ethos of clinical participation is a mixture of ideas, emotions, attitudes and arguments. Together they form and sensitize people's ways of relating to biomedical science in general and clinical trials in particular.

References

Albin, Roger L. 2002: Sham Surgery Controls: Intracerebral Grafting of Fetal Tissue for Parkinson's Disease and Proposed Criteria for use of Sham Surgery Controls. *Journal of Medical Ethics*, 28(5): 322–325.

Barker, Roger A., Jessica Barrett, Sarah L. Mason & Anders Björklund 2013: Fetal Dopaminergic Transplantation Trials and the Future of Neural Grafting in Parkinson's Disease. *Lancet Neurology*, 12(1): 84–91.

Brown, Nik 2003: Hope Against Hype: Accountability in Biopasts, Presents and Futures. *Science Studies*, 16(2): 3–21.

Charmaz, Kathy 1991: *Good Days, Bad Days: The Self in Chronic Illness and Time.* New Brunswick: Rutgers University Press.

Devolder, Katrien 2015: *The Ethics of Embryonic Stem Cell Research.* Oxford: Oxford University Press.

Dunnett, Stephen B., Anders Björklund & Olle Lindvall 2001: Cell Therapy in Parkinson's Disease – Stop or Go? *Nature Reviews*, 2(5): 365–369.

Evans, Jonathan R., Sarah L. Mason & Roger A. Barker 2012: Current Status of Clinical Trials of Neural Transplantation in Parkinson's Disease. *Progress in Brain Research*, 200: 169–198.

Freed, Curt R., Paul E. Greene, Robert E. Breeze, Wei-Yann Tsai, William DuMouchel, Richard Kao, Sandra Dillon, Howard Winfield, Sharon Culver, John Q. Trojanowski,

David Eidelberg & Stanley Fahn 2001: Transplantation of Embryonic Dopamine Neurons for Severe Parkinson's Disease. *New England Journal of Medicine*, 344(10): 710–719.

Freeman, Thomas B., Dorothy E. Vawter, Paul E. Leaverton, James H. Godbold, Robert A. Hauser, Christopher G. Goetz & Warren Olanow 1999: Use of Placebo Surgery in Controlled Trials of a Cellular-based Therapy for Parkinson's Disease. *The New England Journal of Medicine*, 341(13): 988–991.

Gottweis, Herbert 2008: Participation and the New Governance of Life. *BioSocieties*, 3(3): 265–286.

Gottweis, Herbert, Brian Salter & Catherine Waldby 2009: *The Global Politics of Human Embryonic Stem Cell Science: Regenerative Medicine in Transition*. Basingstoke and New York: Palgrave Macmillan.

Hagell, Peter 2004: Utmaningar i utvecklingen: Nya behandlingsmetoder av Parkinsons sjukdom. In: Susanne Lundin (ed.) *En ny kropp: Essäer om medicinska visioner och personliga val*. Lund: Nordic Academic Press.

Idvall, Markus 2017: Synchronizing the Self with Science: How Individuals with Parkinson's Disease Move along with Clinical Trials. *Ethnologia Scandinavica*, 47: 57–77.

Idvall, Markus, Andréa Wiszmeg & Susanne Lundin 2013: *Focus Group Conversations on Clinical Possibilities and Risks within Parkinson Research: A Swedish Case Study*. Lund: Department of Arts and Cultural Sciences, Lund University.

Lundin, Susanne 2004: Etik som praxis. In: Susanne Lundin (ed.) *En ny kropp: Essäer om medicinska visioner och personliga val*. Lund: Nordic Academic Press.

Macklin, Ruth 1999: The Ethical Problems with Sham Surgery in Clinical Research. *The New England Journal of Medicine*, 341(13): 991–996.

Miller, Franklin G. 2003: Sham Surgery: An Ethical Analysis. *The American Journal of Bioethics*, 3(4): 41–48.

Novas, Carlos 2006: The Political Economy of Hope: Patients' Organizations, Science and Biovalue. *BioSocieties*, 1(3): 289–305.

Olanow, C. Warren, Christopher G. Goetz, Jeffrey H. Kordower, A. Jon Stoessl, Vesna Sossi, Mitchell F. Brin, Kathleen M. Shannon, G. Michael Nauert, Daniel P. Perl, James Godbold & Thomas B. Freeman 2003: A Double-blind Controlled Trial of Bilateral Fetal Nigral Transplantation in Parkinson's Disease. *Annals of Neurology*, 54(3): 403–414.

Petit, Geraldine H., Tomas T. Olsson & Patrik Brundin 2014: Review: The Future of Cell Therapies and Brain Repair: Parkinson's Disease Leads the Way. *Neuropathology and Applied Neurobiology*, 40(1): 60–70.

Rubin, Beatrix P. 2008: Therapeutic Promise in the Discourse of Human Embryonic Stem Cell Research. *Science as Culture*, 17(1): 13–27.

Solimeo, Samantha 2009: *With Shaking Hands: Aging with Parkinson's Disease in America's Heartland*. New Brunswick & London: Rutgers University Press.

Wertz, D.C. 2002: Embryo and Stem Cell Research in the United States: History and Politics. *Gene Therapy*, 9(11): 674–678.

Wiszmeg, Andréa 2012: Medical Need, Ethical Scepticism: Clashing Views on the Use of Fetuses in Parkinson's Disease Research. In: Max Liljefors, Susanne Lundin & Andréa Wiszmeg (eds.) *The Atomized Body: The Cultural Life of Stem Cells, Genes and Neurons*. Lund: Nordic Academic Press.

Two afterwords
by Malin Parmar and Aud Sissel Hoel

In a multidisciplinary research collaboration – which this anthology is a result of – it is central to create arenas for dialogue. We have therefore invited two of our colleagues, one from the neuroscience milieu and one from our own research community, to read and freely comment upon our results. In these two afterwords Malin Parmar, Professor of cellular neuroscience and group leader for developmental and regenerative neurobiology at Lund University, and Aud Sissel Hoel, Professor at Norwegian University of Science and Technology, give their perspectives on our results.

Walking around in a minefield of controversies

MALIN PARMAR

I work on translational stem cell biology and as part of my work I am involved in the multicentre foetal cell transplantation trial TRANSEURO study. A large part of my work is focused on developing alternative sources of cells, such as stem cells or reprogrammed neurons. As such, my work on a daily basis involves ethically and societally controversial topics such as use of aborted embryos, human stem cells, animal research. Moreover, in the translational projects carried out in teams, we encounter questions related to trial participation, placebo/sham surgery etc. In other words, I am walking around in a minefield of controversies. So, how do I deal with that? One thing is of course to make sure that my colleagues and I fully comply with ethical guidelines and have appropriate permits from the correct authority for each step we take, where this is required. This may sound easier than it is. Especially since stem cells are just on the verge of hitting clinical trials and a regulatory framework is not always in place. So, we need to work with the regulators to map the road ahead.

I strongly feel, though, that these are not questions that should only be discussed in an academic setting with selected stakeholders. The way we shape our society, develop new therapies, regulate their use and practice new knowledge in moving forward are questions that belong to all of us.

One central focus in this anthology concerns communication between patients, their caregivers and the scientist. This is discussed in chapter 1 (Hansson), 4 (Wiszmeg), 5 (Hansson) and 7 (Idvall), all of which I will return to. The involvement of patients in debates and decision making is not an uncontroversial topic. As discussed in Kristofer Hansson's chapter 'A different kind of engagement: P.C. Jersild's novel *A Living Soul*', the biomedical researcher Dan Graur (2007) raises the issue that it can be unfavourable to include the public in such debate and discussion, since they can have misinformed information and incorrect knowledge. It is a mistake to take this standpoint. We are all part of society. We all have different backgrounds and abilities but we should form our future together. Graur is right in that the general public may not know the facts and enough background to interpret the scientific value. So, one important task for the scientists participating in the public debate is to interpret the facts and describe the relevance to the lay public, regulators and other stakeholders. This will make it easier to have an inclusive discussion and to balance hope versus hype of a particular new therapy and, in return, the patients share their knowledge of what it is like to live with a particular disease, what outcome measures are relevant, what would be a meaningful improvement in terms of quality of life etc. To enable this, however, different groups need to find new arenas and new methods to communicate. Such arenas for communication are described in this anthology with focus on patients and their interaction with scientists, see for example Kristofer Hansson's chapter 'Mixed emotions in the laboratory: When scientific knowledge confronts everyday knowledge' and Markus Idvall's chapter 'Taking part in clinical trials: The therapeutic ethos of patients and public towards experimental cell transplantations'. This is indeed important and I would like to increase the scope to include increased communication within society at large, between scientists from different disciplines, clinicians, politicians and other stakeholders as well. New social media makes it easier to interact. Discussions can take place on internet and both patients and scientists can share their experiences and thoughts via blogs.

Reading this anthology, I cannot help but feeling that scientists are put in a separate category than the rest of society. But we are all part of the same society. We also become affected by diseases, have diverse ethical standpoints, personal opinions etc. And we do not know all the answers. Reading Andréa Wiszmeg's chapter 'Diffractions of the foetal cell suspension: Scientific knowledge and value in laboratory work' about how we as scientists view foetal cells, one gets a little bit of that multifaceted flair. We can reflect on the material that we work with ourselves, and by interacting with different people we can understand different ways of regarding the same material. What is important to add is that perhaps we can view the same thing in different ways in different circumstances. But the day we do the transplant, we need to focus on one thing only: the quality of the cells in the suspension. Are they of a sufficient quality and quantity to transplant to a patient? At that point, anything but viability, quality, cell numbers and phenotype is irrelevant. We are trained to determine what is relevant and to zoom in on specific issues. This training makes us good scientists. But it does not make us good communicators. Discussions amongst ourselves, colleagues and fellow academics, as well as our interactions with patients and society help us understand that the cells in the tube are for some a lost life, and for others a hope for a better life. This helps us in interacting with society at large, in designing new experiments and to move forward in the future.

At first, I find myself slightly provoked by the description of the Swedish author P.C. Jersild's description of the brain in the jar described in chapter one, thinking that this is so wrong (Jersild 1988). Of course you cannot keep a live brain in a jar. The cells that make up the brain would cease to function and die. And even if you could keep the cells alive, the brain would not be able to see, think and feel. So, the suggestion that Ypsilon is violated by the scientists performing experiments becomes absurd. To even suggest this is preposterous. And it puts the scientists in a bad light. Then upon reflection, I am thinking about recent developments in the field of neuroscience and stem cell biology. A few years ago, ground-breaking studies showed that cells in a culture dish can actually self-organize into a three-dimensional structure that contains all the cell types of a specific organ such as the retina or cerebral cortex (Eiraku et al. 2011; Lancaster & Knoblich 2014). Since then, these new ways to culture cells in three-dimensional systems have become increasingly refined, and larger and more complex organoids can be cultured. At a recent meeting, the most advanced organoid culture to date was presented. It contained both neurons and glia (the two main cell types of the brain). The cells formed functioning neural circuits, and sensory neurons in these organoids were shown to respond to visual stimuli (Quadrato et al. 2017). I also recently listened to a podcast about how scientists are creating robots to care for the elderly population in Japan. Part of this work focuses on teaching the robots to understand and respond to human emotions. The robots can, for example, record changes in facial features and voice appearance; they can then select a programme for how to appropriately respond to the signals of the person they are caring for and change their output accordingly.

Although both self-organizing organoids and robots that respond to emotions are miles away from becoming a brain in a dish with thoughts and feelings (Knoblich 2017), or an artificially created electronic brain that responds to sensory input, scientific discoveries inevitably propel us into a future where the impossible may one day become possible (Huch et al. 2017). Collectively, we need to make sure these scientific advancements serve society well and become our tools for a better understanding of diseases and their treatments, better care for the aging population and increase in quality of life without damaging society or putting people at risk. An essential component in this is a continuous and deepened dialogue between society, different stakeholders and regulatory bodies. Furthermore, intensified academic collaborations across different disciplines and fields is absolutely essential.

Therefore, reading this anthology is encouraging. It is an important outcome of a Lund University based research programme that develops new therapies for diseases in the brain. This includes translational gene and stem cell therapies that are expected to reach the clinic in the next few years, as well as more futuristic studies where advanced microfluidics are used to mimic early gradients to create an early foetal brain structure in a dish. Integrated into this research is the cultural research team conducting the valuable studies that are reported in this anthology. Together, we will work for better health. And together, we will steer away from alternative scenarios where the findings are used inappropriately and without benefit for society and patients.

Styles of seeing and knowing in the neurosciences

AUD SISSEL HOEL

In recent years the way we think and talk about ourselves has become increasingly brain-focused, owing no doubt to the rising prestige of the neurosciences – which in turn is made possible by a broad array of new and emerging technologies for mapping and intervening into the living brain. The prospects of the neurosciences to produce insights into human behaviour and to provide cures for neurodegenerative diseases and psychiatric disorders have generated hopes, curiosity, and a strong demand for knowledge in the public. The explosion of interest has spawned what has been referred to as a 'veritable neuro-everything craze' (Pustilnik 2008: 3), ranging from new hybrid disciplines such as neuroethics, neuroeconomics, and neurolaw, to a rich popular media discourse on the most recent neuroscience findings. Researchers in the humanities and social sciences provide critical considerations of both expert and public discourses on brain and mind (Dumit 2004; Joyce 2008; Choudhury & Slaby 2011; Rose & Abi-Rached 2013). The present anthology contributes to these ongoing deliberations, focusing on the participation of non-experts in the discourses on brain and mind. My commentary deals with relations between neuroscience and aesthetics, which is discussed in several of the preceding chapters, more precisely, in chapter 2 (Bengtsen & Suneson), 3 (Liljefors) and 6 (Alftberg & Rosenqvist), all of which I will return to. However, since the approach to aesthetics advocated in this afterword is mediation-centred, I will also touch upon topics addressed in other chapters, such as chapter 1 (Hansson), which, taking its point of departure from a science fiction novel, ponders the implications of a brain-in-a-vat scenario.

'Aesthetics', as understood here, is not restricted to considerations pertaining to artworks or artistic practices. Nor is it limited to questions relating to beauty, such as for example when scientific images, in various kinds of image competitions, are estimated for their aesthetic qualities and prizes are awarded to the most compelling or visually striking entries (a presentation and discussion of such competitions is given in chapter 3 by Liljefors). Revisiting and renewing the older sense of the term 'aesthetics', which concerns sense perception, aesthetics as proposed here is reconfigured as the study of mediation, more precisely, as the study of the way that material apparatuses (bodies, technologies, symbolic systems) regulate the relationship between objects and observers, and hence, the way we see and think. The guiding idea is that knowledge has material bases, in tools and instruments as well as in the active, experiencing body. In this view, the aesthetic and the epistemic are no longer regarded as separate domains but as deeply entangled – not only sometimes, but as a rule. In the following, I will shortly consider three topics of relevance to such a mediation-centred aesthetics: the roles played by metaphors, embodiment and instrumentation in science. I conclude by considering the notion of style, pointing out its pertinence for understanding neuroscientific observation and measurement practices.

In his chapter '"Biospace": Metaphors of space in microbiological images', art historian Max Liljefors discusses the use of metaphors as shorthand explanations. He gives the example of microbiological images, which often present cells, proteins and other biological entities as hovering weightlessly against a dark, empty background. We recognize this mode of presenting from awe-inspiring, popular displays of planets, stars and stellar systems, as exemplified by the iconic image of the Earth commonly referred to as 'The Blue Marble'. Biological entities are presented as if they were celestial bodies, which is why Liljefors coins this space as 'biospace'. He further characterizes this space as purely pictorial, since it is a space that exist solely in and through pictures. The use of metaphors as shorthand explanations is also common in neuroscience. The first example that springs to my mind is the Norwegian neuroscientists and Nobel lauerates May-Britt Moser and Edvard I. Moser, who, when they in popular contexts present their breakthrough discovery of the 'grid cell' (a type of neuron in the brain that allows the organism to keep track of its position in space), refer to this kind of cell as the brain's 'inner GPS'. In both these cases (biological-entity-as-celestial-entity, grid-cell-as-GPS), it is unclear whether the metaphors in question serve a communicative function only, facilitating the understanding of complicated biological or neuroscientific findings, or if they also inform the understanding of the scientific object on the epistemic level. I would say probably both.

In the life sciences, comparisons between biological systems and technical systems are not at all uncommon. When it comes to the neurosciences, there is one metaphor that continues to exercise its influence from the time it was introduced in the 1940s: the brain/mind as an information processor. The computer metaphor of the brain/mind played a key role in the establishment of the research trajectories leading up to the present-day sciences of the brain and mind. It broke new ground by allowing a rigorous study of the elusive mental world through the employment of computer science concepts. Still today the vocabularies for talking about cognitive activities and the workings of the brain are made up almost exclusively by computer terms (hardware, software, input, output, encoding, decoding, processing of information, retrieval of knowledge, storing of memories – the list could go on). Certainly, there is nothing wrong with science making use of metaphors. Quite the contrary, the coining of new, guiding metaphors is an intrinsic part of scientific theory development. It is important to keep in mind, however, that all metaphors, including the computer metaphor, highlight some aspects of the phenomenon of interest at the expense of other aspects, and further, that all metaphors undertake a comparison based on a tentative 'as if'. Even when metaphors are productive in the sense of initiating new lines of research into some phenomena, there is no guarantee that the metaphor is effective in singling out the most pertinent correlations and causal relationships of the said phenomena. The aptness of the computer metaphor of the brain/mind, for example, has been thoroughly questioned all along – most notably, perhaps, by Hubert L. Dreyfus (1972 & 1992). At the same time, and due to its pervasiveness, the computer metaphor has become increasingly naturalized – to the extent that the tentative status of the metaphor is sometimes lost from view, with the effect that, in expert as well as in public discourses, it is often simply presumed that the human brain, in all important respects, behaves like a computer.

The comparison between brains/minds and computers lends support to the idea that mental states, redefined as computational states, can be realized in systems with very different physical constitutions than that of human brains – raising questions such as, for example, whether machines can think. This idea of the multiple realizability of mental states is related to another idea of the relative independence of mental activity from the rest of the human body. These ideas of the separability of the mind from the physical brain, or alternatively, of the separability of the brain from the body, have on the one hand prompted the development of large-scale artificial intelligence research programmes, and on the other hand, lead to a profusion of speculations and fantasies about robots coming alive and developing feelings and emotions. Examples of this include the movie *A.I. Artificial Intelligence* (by Steven Spielberg, 2001); or human beings surviving the amputation of their bodies, either by having their brains kept alive artificially, such as in the novel *A Living Soul* (Jersild 1988), where one of the main characters is a living brain floating in an aquarium (a case discussed by Kristofer Hansson in the first chapter of this anthology); or by having their consciousness uploaded to the Internet, such as in the more recent sci-fi movie *Transcendence* (by Wally Pfister, 2014). Even if most contemporary neuroscientists would never endorse such bizarre views of the brain/mind, neuroscientific approaches tend to be – unsurprisingly, perhaps – rather brain-centred. This is to say that they tend to overlook that cognition is embodied, that is, that cognition is shaped by the involvement of the entire body of the organism, including the organism's movements and interactions with the environment (Varela, Thompson & Rosch 1991; Gallagher 2005). Some proponents of embodied cognition even maintain that cognitive work can be 'off-loaded' onto the environment, making the environment part of the cognitive system (Clark 1997). The proponents of the embodied cognition approach typically challenge the dualist ('Cartesian') under-pinnings of the mainstream sciences of the brain and mind (for a dis-cussion, see Hoel & Carusi 2017). As pointed out by Åsa Alftberg and Johanna Rosenqvist in their chapter dealing with the participation of Alzheimer's patients in art educational situations, mainstream neurosci-entific approaches to selfhood tend to isolate the brain and focus primar-ily on cognitive phenomena, fostering a notion of a 'cerebral subject'. This privileging of cognitive abilities as indications of selfhood, at the expense of a more embodied notion of self, was found by Alftberg and Rosenqvist to underpin even an art-educational initiative directed at dementia-afflicted audiences.

In their chapter 'Pathological creativity: How popular media connect neurological disease and creative practices', Peter Bengtsen and Ellen Suneson point to the way that popular media narratives reinforce sci-entifically inaccurate ideas about creativity as intrinsically linked with neurological diseases. The artistic style of the Dutch painter Vincent van Gogh is an interesting case in point, since, in a BBC-produced popular science programme discussed by Bengtsen and Suneson, van Gogh's style of painting is regarded as indicative of a certain neurological condition in the artist's brain, which is presumed to have been induced by temporal lobe epilepsy. The programme speculates whether the epilepsy might have affected the sensory integration of vision and hearing, resulting in an abnormal sensory experience, which in turn could explain van Gogh's characteristic painterly style. As commented by Bengtsen and Suneson, this line of explanation, which conceives artistic style as a direct reflection of some neurological condition, completely overlooks the historical context of artistic ideas and modes of expression; and in the case of van Gogh, the way that his close collaboration with other artists such as Paul Gauguin and Émile Bernard deeply influenced his artistic style (and vice versa). Following this point, I will add that what is overlooked is that vision has a history – shaped by our habitual interactions with the environment, and just as importantly, by the instruments and tools at our disposal. A circumstance that has been duly pointed out by researchers in art history and visual studies, is that different technologies of vision give rise to different observers and observational practices (Crary 1990). Moreover, and again as hinted at by Bengtsen and Suneson, the kind of explanation that conceives artistic style as a direct reflection of the artist's neurological condition, overlooks that painting is an artistic medium, which itself has a long and multifaceted history. It overlooks that painting, as a medium, has a relative agency and presents the phenomena it targets in accordance with its own norms. As indicated by the term 'style', paintings institute their own mode or manner of displaying the phenomena in question, highlighting some features at the expense of others. This point also holds for the observation and measurement practices of science, which also have their histories and relative autonomy (Daston & Lunbeck 2011). This is also why it makes sense to talk about scientific 'thought collectives' (Fleck 1979 [1935]) or 'epistemic cultures' (Knorr-Cetina 1999). Furthermore, as philosophers of technology never tire of reminding us, scientific vision is also embodied, and to an ever-increasing extent, technologically mediated. In the same manner as metaphors and artistic media, scientific instruments of observation and measurement expose their target phenomena in accordance with their own specific 'amplification-reduction structure' (Ihde 1979: 21).

Tying this point regarding the amplification-reduction structure of mediating apparatuses back to Liljefors's idea of a space of exposure that emerges and exists, in his case, only in and through a micrographical mode of picturing, throws new light on the relationship between the aesthetic and the epistemic. Each mediating apparatus opens a space that is at once aesthetic and epistemic: a specific mode of exposing – displaying and expounding – the target phenomena. As in the case of Liljefors's biospace, the spaces opened up by the various apparatuses emerge and exist only in and through their respective apparatuses. It would make sense, therefore, to talk about 'styles' of seeing and knowing in the neurosciences, which again, would call for the need to analyse the observation and measurement apparatuses of these sciences (including metaphors) with a view to laying bare their specific amplification-reduction structures.[1]

---

1. Thanks to Asle H. Kiran and Annamaria Carusi for their helpful comments on a previous version of this afterword.

References

Choudhury, Suparna & Jan Slaby 2011: *Critical Neuroscience: A Handbook of the Social and Cultural Contexts of Neuroscience*. Chichester: Wiley-Blackwell.

Clark, Andy 1997: *Being There: Putting Brain, Body, and World Together Again*. Cambridge: MIT Press.

Crary, Jonathan 1990: *Techniques of the Observer: On Vision and Modernity in the Nineteenth Century*. Cambridge: MIT Press.

Daston, Lorraine & Elizabeth Lunbeck 2011: *Histories of Scientific Observation*. Chicago and London: University of Chicago Press.

Dreyfus, Hubert 1972: *What Computers Can't Do*. New York: MIT Press.

Dreyfus, Hubert 1992: *What Computers Still Can't Do: A Critique of Artificial Reason*. New York: MIT Press.

Dumit, Joseph 2004: *Picturing Personhood: Brain Scans and Diagnostic Identity*. Princeton: Princeton University Press.

Eiraku, Mototsugu, Nozomu Takata, Hiroki Ishibashi, Masako Kawada, Eriko Sakakura, Satoru Okuda, Kiyotoshi Sekiguchi, Taiji Adachi & Yoshiki Sasai 2011: Self-organizing Optic-cup Morphogenesis in Three-dimensional Culture. *Nature*, 472(7341): 51–56.

Fleck, Ludwik 1979 [1935]: *Genesis and Development of a Scientific Fact*. Chicago: University of Chicago Press.

Gallagher, Shaun 2005: *How the Body Shapes the Mind*. Oxford: Clarendon Press.

Graur, Dan 2007: Public Control Could be a Nightmare for Researchers. *Nature*, 450: 1156.

Hoel, Aud Sissel & Annamaria Carusi 2017: Merleau-Ponty and the Measuring Body. *Theory, Culture & Society*. First published date: 7 February 2017.

Huch, Meritxell, Juergen A. Knoblich, Matthias P. Lutolf & Alfonso Martinez-Arias 2017: The Hope and the Hype of Organoid Research. *Development*, 144(6): 938–941.

Ihde, Don 1979: *Technics and Praxis*. Dordrecht: Reidel.

Jersild, P.C. 1988: *A Living Soul*. Norwich: Norvik Press.

Joyce, Kelly A. 2008: *Magnetic Appeal: MRI and the Myth of Transparency*. Ithaca & London: Cornell University Press.

Knoblich, Juergen A. 2017: Lab-Built Brains. *Scientific American*, 316: 26–31.

Knorr-Cetina, Karin 1999: *Epistemic Cultures: How the Sciences Make Knowledge*. Cambridge: Harvard University Press.

Lancaster, Madeline A. & Juergen A. Knoblich 2014: Organogenesis in a Dish: Modeling Development and Disease Using Organoid Technologies. *Science*, 345(6194).

Pustilnik, Amanda C. 2008: *Violence on the Brain: A Critique of Neuroscience in Criminal Law*. Harvard Law School Faculty Scholarship Series, paper 14.

Quadrato, Giorgia, Tuan Nguyen, Evan Z. Macosko, John L. Sherwood, Sung Min Yang, Daniel R. Berger, Natalie Maria, Jorg Scholvin, Melissa Goldman, Justin P. Kinney, Edward S. Boyden, Jeff W. Lichtman, Ziv M. Williams, Steven A. McCarroll & Paola Arlotta 2017: Cell Diversity and Network Dynamics in Photosensitive Human Brain Organoids. *Nature*, 545(7652): 48–53.

Rose, Nikolas & Joelle M. Abi-Rached 2013: *Neuro: The New Brain Sciences and the Management of the Mind*. Princeton: Princeton University Press.

Varela, Francisco J., Evan Thompson & Eleanor Rosch 1991: *The Embodied Mind: Cognitive Science and Human Experience*. Cambridge: MIT Press.

Contributors

ÅSA ALFTBERG, PhD in ethnology, works at the Department of Arts and Cultural Sciences, Lund University and the Department of Social Work, Malmö University. Her research concerns body, health and materiality, for instance experiences of the ageing body as well as experiences of hearing loss and how meaning is created and negotiated in the intersection of body, technology and everyday life. Currently she focuses on cultural narratives of the brain.

PETER BENGTSEN is an art historian and sociologist working as assistant professor at the Department of Arts and Cultural Sciences, Lund University. His research interests include street art, graffiti, the publicness of urban public space, spatial justice, and the representation of neuroscience and neurological disease in popular media.

KRISTOFER HANSSON is associate professor of ethnology and researcher at the Department of Arts and Cultural Sciences, Lund University. He did his PhD studies at Vårdalinstitutet – The Swedish Institute for Health Sciences. His research focus is cultural analysis of medical praxis in health care and biomedical research. In recent years much of his research is related to citizen participation in new biomedical technologies.

AUD SISSEL HOEL is professor of media studies and visual culture at the Norwegian University of Science and Technology. Her research concerns the roles of images and tools in knowledge and thinking, focusing on photography, science images, scientific instruments, medical imaging and visualisation. Hoel's publications cover a wide range of topics on the overlapping fields of visual studies, science studies and media philosophy.

MARKUS IDVALL is associate professor of ethnology and senior lecturer at the Department of Arts and Cultural Sciences, Lund University. His research has focused on medical and health care issues connected with diabetes and renal failure for some years. In a current project, he is interested in how clinical research on Parkinson's disease has an impact on the social and cultural relationship between scientists and patients.

MAX LILJEFORS is professor of art history and visual studies at the Department of Arts and Cultural Sciences, Lund University. He has published extensively on topics such as Holocaust representation, visual historiography, medical visual cultures, law and art, as well as video and performance art. His current research is focused on aesthetic experiences in relation to health and spirituality.

MALIN PARMAR is a professor in cellular neuroscience at Lund University and at Robertson investigator – a New York Stem Cell Foundation. Her research topic is translational stem cell biology and brain repair, with particular focus on developing new cell based therapies for Parkinson's disease.

JOHANNA ROSENQVIST is senior lecturer in history and theory of craft at Konstfack, besides her tenure as senior lecturer in art history and visual studies at Linnaeus University. Her research concerns performativity in arts, crafts and design and focuses on the aspects of making art in relation to a wider field of cultural production. She has co-edited an anthology about the art educational practice of museums and she has conducted research projects developing methods for art and making art as a tool for therapeutic rehabilitation for people with neurodegenerative diseases.

ELLEN SUNESON is a PhD candidate in art history and visual studies at the Department of Arts and Cultural Sciences, Lund University. Her research interests concern queer feminism, performativity and art historiography. Suneson has been part of various projects concerning art and neuroscience – with focus both on the intersection between popular culture and neuroscience and on developing methods for art therapy aimed at Parkinson's patients.

ANDRÉA WISZMEG is a PhD candidate in ethnology at the Department of Arts and Cultural Sciences, Lund University. Her research focus is various social, cultural and material dimensions of biomedical research and practice, and their relation to science theory and to research ideals.

Previously published in the Pandora Series

Boel Berner, *Perpetuum mobile? Teknikens utmaningar och historiens gång* (1999)

Johan M. Sanne, *Creating Safety in Air Traffic Control* (1999)

Sabrina Thelander, *Tillbaka till livet. Att skapa säkerhet i hjärtintensivvården* (2001)

Kerstin Sandell, *Att (åter)skapa "det normala". Bröstoperationer och brännskador i plastikkirurgisk praktik* (2001)

Viveka Adelswärd & Lisbeth Sachs, *Framtida skuggor. Samtal om risk, prevention och den genetiska familjen* (2002)

Boel Berner (ed.), *Vem tillhör tekniken? Kunskap och kön i teknikens värld* (2003)

Christer Eldh, *Den riskfyllda gemenskapen. Att hantera säkerhet på ett passagerarfartyg* (2004)

Ericka Johnson, *Situating Simulators. The Integration of Simulations in Medical Practice* (2004)*

Tora Holmberg, *Vetenskap på gränsen* (2005)

Staffan Wennerholm, *Framtidsskaparna. Vetenskapens ungdomskultur vid svenska läroverk 1930–1970* (2005)

Elin Bommenel, *Sockerförsöket. Kariesexperimenten 1943–1960 på Vipeholms sjukhus för sinneslöa* (2006)

Robert Hrelja, *I hettan från ångpannan. Vetenskap och politik i konflikter om tekniska anläggningar* (2006)

Jonas Anshelm, *Bergsäkert eller våghalsigt? Frågan om kärnavfallets hantering i det offentliga samtalet i Sverige 1950–2002* (2006)

Christer Nordlund, *Hormoner för livet. Endokrinologin, läkemedelsindustrin och drömmen om ett botemedel mot sterilitet 1930–1970* (2008)

Mats Brusman, Tora Friberg & Jane Summerton (eds.), *Resande, planering, makt* (2008)

Johan Wänström, *Samråd om Ostlänken. Raka spåret mot en bättre demokrati?* (2009)

Ingemar Bohlin & Morten Sager (eds.), *Evidensens många ansikten. Evidensbaserad praktik i praktiken* (2011)

Maria Björkman, *Den anfrätta stammen. Nils von Hofsten, eugeniken och steriliseringarna 1909–1963* (2011)

Annemarie Mol, *Omsorgens logik. Aktiva patienter och valfrihetens gränser* (2011)

Boel Berner, *Blodflöden. Blodgivning och blodtransfusion i det svenska samhället* (2012)

Olof Hallonsten (ed.), *In Pursuit of a Promise. Perspectives on the Political Process to Establish the European Spallation Source (ESS) in Lund, Sweden* (2012)*

Boel Berner & Isabelle Dussauge (eds.), *Kön, kropp, materialitet. Perspektiv från fransk genusforskning* (2014)

Lisa Lindén, *Communicating Care. The Contradictions of HPV Vaccination Campaigns* (2016)*

Anders Lundgren, *Kunskap och kemisk industri i 1800-talets Sverige* (2017)

---

* Also available in Arkiv Academic Press international editions, please visit www.arkivacademicpress.com for up-to-date information on available titles.

Arkiv Academic Press

Arkiv Academic Press is an imprint of the Swedish publishing house Arkiv förlag. For up-to-date information on distribution and available titles, please visit:

www.arkivacademicpress.com

Published books

Ericka Johnson, *Situating Simulators. The Integration of Simulations in Medical Practice* (paperback 2012 [original edition by Arkiv förlag 2004])

Olof Hallonsten (ed.), *In Pursuit of a Promise. Perspectives on the Political Process to Establish the European Spallation Source (ESS) in Lund, Sweden* (paperback 2012)

Rebecca Selberg, *Femininity at Work. Gender, Labour, and Changing Relations of Power in a Swedish Hospital* (paperback 2012)

Sven E O Hort (birth name Olsson), *Social Policy, Welfare State, and Civil Society in Sweden.* Volume I: *History, Policies, and Institutions 1884–1988* (hardcover & paperback 2014, 3rd enlarged edition [1st edition by Arkiv förlag 1990])

Sven E O Hort (birth name Olsson), *Social Policy, Welfare State, and Civil Society in Sweden.* Volume II: *The Lost World of Social Democracy 1988–2015* (hardcover & paperback 2014, 3rd enlarged edition [1st edition by Arkiv förlag 1990])

Lisa Lindén, *Communicating Care. The Contradictions of HPV Vaccination Campaigns* (paperback 2016)

Gunnar Olofsson & Sven Hort (eds.), *Class, Sex and Revolutions. Göran Therborn – A Critical Appraisal* (paperback 2016)

Kristofer Hansson & Markus Idvall (eds.), *Interpreting the Brain in Society. Cultural Reflections on Neuroscientific Practices* (paperback 2017)

www.ingramcontent.com/pod-product-compliance
Lightning Source LLC
Chambersburg PA
CBHW050805270326
41926CB00025B/4549